AF599236

and the Decline of the Family Dairy Farm

Thomas Sieger

Little Creek Press®
A Division of Kristin Mitchell Design, Inc.
5341 Sunny Ridge Road
Mineral Point, Wisconsin 53565

Book Design and Project Coordination:
Little Creek Press and Book Design

Cover Art: Arcane Book Covers
Cover Photo: © 2020 John P. Gauder

First Printing
November 2021

Printed in the United States of America

For more information or to order books,
www.littlecreekpress.com

Library of Congress Control Number: 2021921576

ISBN-13: 978-1-955656-12-2

Note: The characters in this book are fictional. The setting of Woodstock and its environs are in Richland County, Wisconsin.

Dedicated to the farm families who have been thoughtful stewards of their dairy herds and driftless lands.

"

The cow is the foster mother of the human race.

"

Former Wisconsin Governor
William D. Hoard,
one of the founders of the
Wisconsin Dairymen's Association

Table of Contents

Sentinels of the Heartland

Undulating and verdant
rising out of the glaciated plain
interrupting the familiarity of the prairie
shadowing the fertile coulees
towering over the great river valley.

Rural and wild
crowned by sandstone cliffs
carpeted by lush hardwood forest
drained by clear springs and streams
carved by rural by-ways.

Home and haven
refuge of the family farm
retreat and inspiration of artisans and dreamers
remnant of native lands
siren of childhood memories.

The bluffs of the Driftless—calling me home,
calling me home.

—Tom Sieger

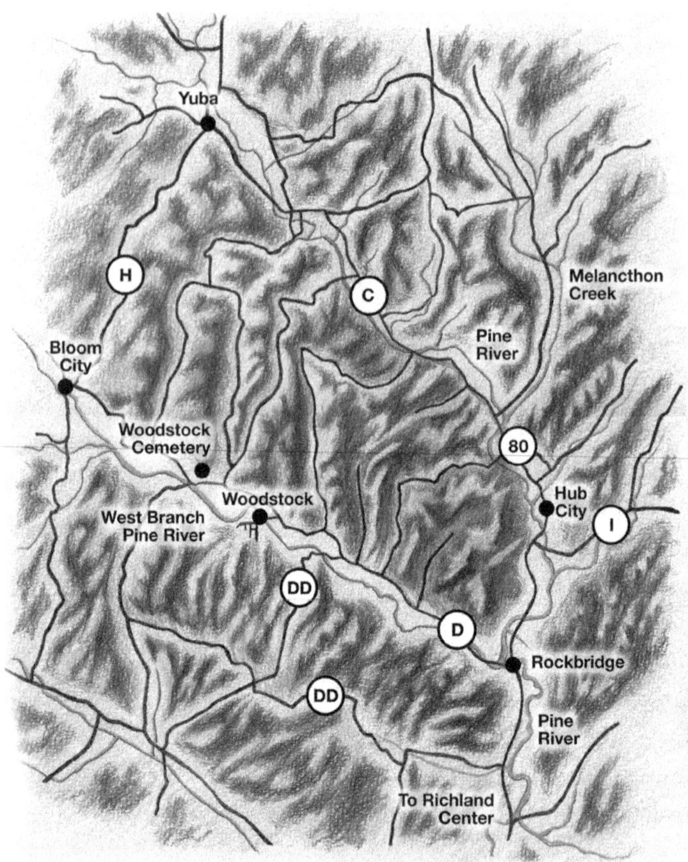

Upper Pine River Valley, Richland County, Wisconsin.
Map rendered by Robin Good.

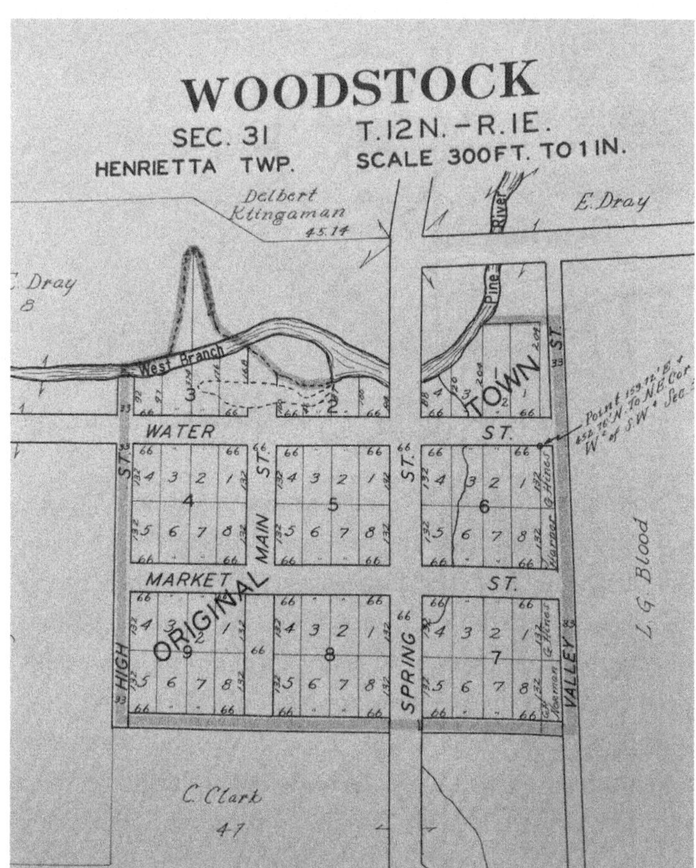

Early map of Woodstock, Wisconsin. Atlas of Richland County, Wisconsin. Rockford, IL: General Map Company, 1951.

Denny Schulz

Coming back from a hardware run to Farmers Supply, Denny Schulz swung his worn F-150 pickup left on County D into the West Pine River Valley. With a strong physique, weathered skin, and calloused hands signaling the strenuous occupation of dairy farming, he slowed his speed to take in the full panorama of the valley. "God, how I love this place," he thought. It was a love that seemed to grow stronger as he approached his sixtieth birthday. With gentle, heavily wooded bluffs framing each side of the valley and the clear, meandering West Branch of the Pine River coursing down the center, he never tired of the feeling of coming home. Memories flooded over him as he snaked north up the valley toward the turnoff to his farm. He looked over his right shoulder to see the well-groomed lawn of the Woodstock Cemetery, the resting place of his aged mother, who recently had joined his father, grandparents, and great-grandparents. "Hell," he thought, "I bet I've got thirty relatives planted in that place." He made a mental note to remind his wife, Mary, to get some flowers ready for the upcoming Memorial Day celebration.

Over his left shoulder, along the river, were the ragged remains of the town of Woodstock. Although this is where it all started with the Schulz family, nothing much was left today of the once-bustling frontier village. Even the original family log home had succumbed long ago to fire and flooding. It was now a largely abandoned crossroads of just a few individuals living a rural lifestyle. Was it just his imagination, or was he getting overly sentimental in his old age as the crushing debt of the farm seemed to take a tighter grip on his throat? He was fighting hard to keep his ever growing fear of the future hidden beneath the veneer of a rugged, self-reliant Wisconsin dairyman. "Buck up, buddy," he thought, taking a left, crossing the river, and heading onto Schulz Lane. "Somehow we will get through this. For Mary and the kids, we'll have to." As he passed the sign marking the Schulz Family Century Farm, he heard it again—a slight perceptible whine coming from the rear end—a failing differential or bearing problem? "Oh, shit," he thought at the prospect of a pickup repair bill on top of the pile of mounting debt.

The West Pine River Valley

The West Branch of the Pine River zig-zags down one of the prettiest valleys you will find in the Midwest's Driftless area, a four-state region bisected by the Mississippi River. The Driftless region was spared from the glacial scouring 10,000 years ago during the Pleistocene Epoch. Although the exact boundaries of the region are open to geological interpretations, it covers only a part of four states: southwestern Wisconsin, southeastern Minnesota, northwestern Illinois, and northeastern Iowa, including about 57 counties and 24,000 square miles. In the interest of avoiding a long and tedious explanation of the confusing term "Driftless," suffice to say that you will immediately recognize the area when you find yourself in it. Coming out of the relatively featureless landscapes of the great prairies of the Midwest, here you will find lush green hills, high bluffs, and winding deep valleys, each drained by clear springs and streams that flow to small rivers emptying into the mighty Mississippi, which, some say, contains the greatest concentration of clear, cool running streams in the world.

Although the white settlers of European descent had the greatest impact on the Driftless area, they were not the earliest to inhabit these beautiful bluffs and valleys. Archeological records, primarily unearthed tools and artifacts, suggest that the nomadic Mound Builders in the fifteenth and sixteenth centuries were some of the first peoples to inhabit the Driftless areas. Interestingly, implements made of copper have been found from the mound-building period, suggesting trade connections with the Lake Superior area as no copper ores are found in Richland County. Prior to the arrival of the first white settlers, the area was home to the Ho-Chunk Nation and a lesser number from the Potawatomi Tribe. The Ho-Chunk called the high bluffs and hills of the Driftless area a name phonetically like "Ocooch," and today you will see references to the Ocooch Mountains on some of the earlier maps and recorded history of the region. Early records indicated very little strife between the white settler and the Indians, one record stating:

> Their intercourse with the settlers came to be regarded more as an occasion of pleasant remembrances than of dread or danger. Some pleasant friendships were formed between the pioneer family and the former owners of the land which the paleface was tilling.[1]

However, in one of many sad chapters of U.S. government relations with our indigenous peoples, the Ho-Chunk Nation was coerced into an 1837 treaty meeting with the federal government in Washington. Under duress, they ceded all

---

1 History of Richland County Wisconsin. 1906. Edited by Judge James H. Miner. Western Historical Association.

their lands east of the Mississippi to the federal government. In turn, the Ho-Chunk Nation agreed to move west within eight months to reservation lands near Omaha, Nebraska, which previously had been conveyed to them by treaty.

Woodstock, Richland County, Wisconsin

The first white settlers arrived at the town of Woodstock in 1853. Some moved from farms in Vermont and points east, hoping to find better land and more opportunity in the Midwest. Others were already in the Midwest, many moving north from working the lead mines around Galena, Illinois, and what is now Lafayette County, Wisconsin. What was the attraction? Initially, the pine and hardwood timber resources were plentiful in the West Pine River Valley. This resource and the availability of waterpower from the Pine River held the potential for mills to process the lumber. As these first settlers set about the work of felling timber in the valley, stories sprang up of log rafts channeling down the pine to sawmills located on the Wisconsin River to the south. Soon, however, Woodstock had its own sawmill.

The village was laid out as required by law in the spring of 1855 along the banks of the West Pine River. Milton Satterlee and Quinton Nicks were responsible for drawing up the layout. In a letter at the time, Milton stated:

> Before laying out the village, Quinton Nicks and I proceeded to clear off what is now the village plot, and the surrounding lands were wild woods. Alder and thorn trees, and under brush [sic] of almost every kind, among the large elms and oak, made the spot where the village now stands one dense forest, more fit for wild prowling beasts than for civilization ... the West Branch and tributaries literally swarmed with speckled trout. Boys and men often caught them with their hands in the spring branches.[2]

Soon after, Woodstock became a bustling village, boasting a post office, a log school, two grist mills, two blacksmith shops, two saloons, two churches, three stores, and a cheese factory. During the decade of 1850 to 1860, the population of Richland County, which includes Woodstock, swelled from less than 1,000 to 9,732.

The first recorded death in Woodstock was the infant child of a Dr. Byers in the fall of 1856. This sad milestone prompted the siting of the Woodstock Cemetery across the river and on a gentle west-facing slope across from town. The Woodstock Cemetery would soon house more residents than lived in the town, with the well-tended landscape standing in direct contrast to some of the deteriorating conditions remaining in the community.

In the history of Richland County, Milton Saterlee indicated that no matter where the settlers came from,

---

2 "Milton Satterlee Tells About Laying Out of Woodstock." 9 February 1961. Republican Observer. Richland County History Room. Brewer Public Library. Richland Center, Wisconsin."

> Mutual desires and interests made them all akin, and by a silent process of "benevolent assimilation" they were converted in a Richland County Family. Among them, there existed very little distinction in worldly circumstances and modes of life—the disparities in conditions that we now observe having been developed gradually in the country and emphasized by the frowns and smiles of that giddy dame, Fortune. It was neither the indolent nor the opulent, as a general fact, who sought homes in this region, for none but industrious men of moderate means would care to endure the preliminary privations and encounter the dangers that they knew would attend them while building homes in the almost unbroken wilderness. They came to better their condition in life by becoming landowners instead of tenants, to rid themselves of a species of landlordism that prevailed in the eastern states, and to emancipate themselves from a condition of semi-vassalage which threatened a doom of servitude for themselves and their children.[3]

Among the early settlers of Woodstock were young Douglas Schulz and his wife, Henrietta, a couple of German descent who moved west from a farm in the thin, rocky soils of Vermont to the promise of more fertile lands in Woodstock. Douglas initially found work in the sawmill, eventually purchased 120 acres adjacent to the north side of the village, and began the arduous task of clearing the land and looking forward to the prospect of wheat farming, which was the main crop of farms in the area during the mid-1800s.

---

3 History of Richland County Wisconsin. 1906. Edited by Judge James H. Miner. Western Historical Association.

But in a reoccurring theme that will play out numerous times in our story, the vagaries of work and farming were to become a fact of life for the village of Woodstock and the West Pine River Valley. As the more level terrain of the river bottoms and gentle slopes of the hillsides were denuded of timber, first came the demise of the lumbering industry in the valley. Initially, Douglas and his family turned to wheat farming on the recently cleared land. However, with annual crops taking a toll on the soil, leading to decreasing annual yields, paltry prices for a bushel of wheat and the opening of land to the west more suitable to grain production, it quickly became evident that this part of Wisconsin was never to be a major producer of grains for kitchen tables of this country. Word began to get around the valley that farmers to the east, in the rich Sauk Prairie area, were becoming successfully established in dairy production. One farmer, in particular, N.W. Morley of neighboring Sauk County was said to have initiated the change leading Wisconsin to become "America's Dairyland." By the 1930s, Wisconsin had around 150,000 small dairy farms, the largest number of any state in the nation.

Douglas cautiously began a small dairying operation, milking by hand a herd of around a dozen cows that were pastured on the bluff sides of the farm and housed during the brutal winters in a rough straw shed. Grains were now primarily grown for animal feed, and the rich cow manure was applied back to the land to increase the fertility of pasture and cropland. Refrigeration for the milk pails was provided by a spring house cooled by the constant forty-five-degree water of one of many springs in the area feeding the Pine River. Fluid milk was mainly sold locally to the local cheese factory to produce cheese and butter, which, in turn, was sold around the county.

As Douglas and Henrietta's family grew, so did the success of the dairying operation. Eventually, the mortgage was paid off, and the old log house—subject to flooding and lost in a tragic fire—was replaced by a handsome brick structure on higher ground. The primitive straw shed was replaced by one of the iconic red Wisconsin dairy barns being promoted by the University of Wisconsin School of Agriculture at the turn of the twentieth century. It was an elongated structure with a high gambrel roof to provide hay storage, built into the side of the hill to provide direct access to the haymow. The lower level housed stanchions for forty dairy cows on a solid concrete floor for improved sanitation and cleanliness. And, most important of all, was the building of an upright silo for the storage for corn, the preferred grain among dairy farmers, which would provide winter feed to allow milking to continue during the harsh Wisconsin winters.

The turn of the century witnessed great change and saw the valley and surrounding Richland County transition to a successful dairy hub. However, these changes also marked the beginning of the demise of our village of Woodstock. In the words of Joseph Kelly, born and raised near Woodstock:

> The growing of wheat in Wisconsin was discontinued, so no use for flouring mills, timber was exhausted ... no need for sawmills, good roads made it possible for trucks to quickly carry the milk of this rich dairy area to the Carnation plant and creameries at Richland Center, so the cheese factory went away.[4]

---

4 I. 1884. Joseph Kelly. Richland Center History Room. Brewer Public Library.

Within one generation, Woodstock experienced its boom days and began its demise. Joseph Kelly went on to write:

> It seems strange indeed that an inland village in a good farming area ... became but a ghost-town, with nothing more to awaken and arouse the interests of its citizenry, than the ringing of the church bell on Sunday morning, and the only spark of the industry of former years remaining, being a grocery store, which frequently changes hands.[5]

Woodstock, Wisconsin was never destined to see the notoriety of its similarly named cousin, Woodstock, New York – famous for the three days of "Peace and Music" in 1969 known as the Woodstock Music Festival. Outside of the few families living in northern Richland County, few if any Wisconsin residents would even know that a village by the name of Woodstock even existed within the State.

However, the dairy farm of Douglas and Henrietta provided a livelihood and source of pride for the subsequent four generations of the Schulz family living in this beautiful valley. But as the old saying goes, "The only thing inevitable in life is change," and the farm economy and dairy industry soon experienced great change once again.

---

5 Early History of Woodstock, Wisconsin. 1884, Supplement 1956. Joseph Kelly. Richland Center History Room. Brewer Public Library. Richland Center, Wisconsin.

A log home built in 1884, reminiscent of homes built by early settlers to the Upper Pine River Valley, was moved from Yuba to the Peckham farm. Copyright 2020 John P. Gauder.

The Ross Gillingham homestead, built in the late 1800s and located just west of Woodstock, is an example of the more substantial structures built by farmers experiencing early success in the dairy industry. Copyright 2020 John P. Gauder.

Storm Clouds Gathering

As Denny drove his pickup down the gravel road into the farmyard, he immediately noticed the gleaming late-model Chevy Silverado parked in front of the house. He recognized the logo on the side door of the local feed co-op. "Oh, shit," Denny thought, "Here we go. Buckle up, bucko."

Putting on his best forced smile, he jumped out of his truck and offered a pleasant greeting and handshake to his visitor. "Hello, Randy. What brings you out to the boonies of Woodstock?"

Randy, the well-dressed older co-op agent looked down at his boots, appearing uncomfortable in the message he was about to deliver. Starting on a pleasant note, he said, "Hello, Denny. Nice to see the start of a good corn crop in your fields along your lane. With some good weather and, God forbid, no flooding like last summer, looks like you'll have full silos come fall."

Hoping to interrupt the delivery of bad news, Denny asked, "Randy, can I get you a brewski?"

"No, I'm on the clock and, besides, I think you know why I'm here. We've done business together for a long time. Hell, way before you were a twinkle in your daddy's eye, we've supplied your farm with feed and fertilizer, and we've come through some tough times together. But, damn it, Denny, we just can't float you any longer. Right now, you're over your head to the tune of $25,000. The manager is breathing down my neck, and I promised I would come out and see if we could arrange some payment."

"Come on, Randy, you know my story as well as the few of us that are left trying to keep our farms afloat in this valley." As he angrily kicked a stone down the driveway, he exclaimed, "With milk prices being what they are, you know we can't even cover our operating expenses, let alone make a simple living! Hell, if Mary hadn't taken the job at the nursing home, we wouldn't even be able to keep the wolf away from our door."

"Denny, I know you and other farmers in the valley are in a really tight spot. I haven't seen times this bad since the crash of 2008. I am sorry, but I gotta tell you the manager is considering freezing any additional product deliveries and making a referral to a collection agency. You know as well as I that this won't end well. Have you considered visiting Tom at the bank in town and refinancing your operating loan?"

"Dream on, Randy! They will probably lock the doors if they see me coming down the street. How many times do you think I can refinance without saddling the kids with a loan that will live on well beyond Mary and me? Hell, I am already up to my Adam's apple in debt!"

"I hate to say this, but we are going to need an answer or some kind of payment plan by the end of the month. Give things some thought and give me a call by Memorial Day. And, Denny, hang in there. We are all rooting for you! I've been there. Since Glenda and I sold the farm, and I took the job at the co-op, we've made a pretty good life for ourselves in Center. Have you considered this option?"

Looking closely as Randy, Denny noticed the web of spider veins on his ruddy cheeks and nose, suggesting that perhaps an abundance of alcohol was helping the difficult transition from the farm to city life. "Randy, five generations of Schulz's have lived on this land, and I'll be damned if I'm gonna be the one to lose the ranch. Besides, cows and farming are all I know. What would I do in town? With this crappy farm economy, the only jobs left are barely minimum wage at the food processing plant. Look at our town. Empty store fronts on Main Street, homes and buildings haven't been maintained for years, and our brightest and best young ones—including my kids—are heading for the hills. Hell, even if I did sell, who would wanna buy this place? This land won't support a larger dairy herd, and I'll be damned if I'm going to sell to some asshole from Chicago for $3,000 an acre so he can have a hunting getaway!"

Since there wasn't much else that could be said, Randy abruptly got into his pickup and rolled down the window, saying, "Give my best to Mary and the kids when you talk to them. We'll be in touch." The truck left a trail of dust down the gravel lane.

Denny's mind reeled as he thought of the implications of the last few minutes. After a couple of years of unusually heavy rainstorms and serious river flooding, he was still short on hay and feed for the cows. God forbid if the co-op cut off his feed supplements. Weighing even heavier on his mind was the moment of reckoning with his wife, Mary. He hadn't exactly been truthful about the load of debt the farm was carrying and knew the time was soon to come for a heart-to-heart conversation. But, thankfully, Randy had paid his visit late in the afternoon while Mary was working the second shift at the nursing home, giving him a bit more time to think if there might be a way out of this shitstorm.

Denny headed back to the old house, and once inside, he immediately stopped to look around. He grew up in this place, and it hadn't changed much since he was a kid. A two-story, well-built structure of brick (reportedly fired right here in Woodstock and gracefully designed), it needed some maintenance. The exterior brick needed tuck pointing, the wide board floors could use refinishing, and the parlor badly needed fresh paint. The woodwork and trim, made of walnut most likely taken from the adjacent hills and milled in the village, needed varnish. Although a bit rough by today's millwork standards, it lent a depth of beauty and character simply not found in more recent structures. Also, the worn but comfortable furnishings and family antiques provided a warm and welcoming sense of home. It often housed three generations of the Schulz family – all playing vital roles in the operation of the family farming enterprise. "Here I go again," he thought, as nostalgia and memories engulfed him.

Heading to the kitchen, he opened the fridge to grab a Spotted Cow Lager but first saw the Tupperware of chicken vegetable soup Mary left him for dinner. "Jeez. She needed to get off this health kick." He could really use a good piece of beef. He decided to pass on the beer as the situation called for something stronger. Rummaging around in the cupboard above the fridge, he found the bottle of bourbon and proceeded to make himself a long Jack and Coke.

Heading back to the front porch, he took a seat on the porch swing and slowly looked up and down the valley. The homesite on higher ground up the side of the western slope confirmed the early lessons from river flooding and provided a great view of the valley and the remains of the town of Woodstock to the south. How he loved nothing more than the light green hues of early spring on the hillsides as the hardwoods were leafing out. Their color was even more intense as the sun began to set behind the western bluff. A pair of sandhills trumpeted as they gracefully flew up the valley, a welcome sound he had missed over the winter. Looking over at the barnyard, he saw his girls slowly coming down the hill, knowing it would soon be time for the evening milking.

Taking a long slow draw on his Jack and Coke, he thought about the options before him to either save the farm or cash out of the dairy business altogether. Selling his milking herd would bring in a little north of a hundred grand, and his thirty heifers about twenty grand—enough to pay off his operating loan. Leasing his forty acres of tillable land for corn production would provide some modest annual income, about a hundred dollars per acre, but he would still have to find work in town to pay off his capital loan and provide a life for him and Mary, something he really struggled with. "The long and short of it is that I am a Wisconsin dairyman at heart. That's all I know and love," he thought. As he looked over the Century family farm from the porch, the rich green pastures on the hillside, and the deep black soils of his cornfields, it was an easy decision. He would go to the bank tomorrow—a long shot—but the last best hope he had to save the farm. Plan B would be a talk with a fellow dairyman down the valley looking to do a major expansion to his herd and dairy business.

Checking his watch, he saw that he just had time to finish his drink, call the bank to set up an appointment for tomorrow, and change into his barn clothes before the hired hand Sam arrived to help with the evening chores.

The Hail Mary Pass

At three o'clock the next afternoon, immediately after Mary left for her shift at the nursing home, Denny was in his pickup driving south down the valley for his three-thirty meeting with Tom at the Bank of Richland Center. Although he did not hold out much hope for any relief, he knew that he had to make this last-ditch effort to save the farm. Saving it for whom, he wondered. He and Mary had reconciled themselves to the fact that neither their son nor daughter would ever live – much less work – on their land again. Both kids seem to relish returning home for family time on weekends, holidays and special occasions and paid lip service to never allowing the farm to pass out of the family's hands. But his oldest daughter Kate had gotten a two-year nursing degree at UW Center, had gone on to complete her bachelor's degree in nursing at UW–Madison, and was now married, looking to start a family, and working in Madison at University Hospital.

He initially held out hope that his son Charlie would continue the long line of Schulz's to farm this land. Charlie had taken an immediate love to the Driftless. From an early age, he eagerly enjoyed trout fishing on the West Pine, the family tradition of the fall deer hunt around Thanksgiving, taking long hikes in the bluffs behind the family home, and morel mushroom hunting in the spring. But it was soon evident that he hadn't developed a love for the cows and the time in the barn that was so essential for a future dairyman. There was no degree in Ag in the future for this boy. Instead, he studied physical education at UW–La Crosse and, like so many other millennials, was off the farm to teach in a La Crosse high school while chasing girls and pursuing his love of fishing and the outdoors on the Mississippi River and in the nearby bluffs. Given the demands of twice a day milking, backbreaking work, and the lack of financial rewards, who could blame them? Denny knew deep down that he was trying to keep things afloat for himself. Farming was the only thing he knew. He loved the land and the animals and certainly did not want to be the Schulz who couldn't keep the farm afloat.

The abandoned dairy barns and ragged homesteads on either side of County Trunk D told the story of the change that had come to the valley over the last couple of decades. As was the case with his own kids, young adults moved off the farm and into the cities as soon as they were able, leaving behind an aging population of farmers unable to make a decent living in the dairy industry. Sadly, the handwriting was on the wall – Denny and Mary were one of the last dairy farms trying to make a go of it in the valley. Was this a fool's errand to approach the bank today?

At the last minute, Denny decided to turn off County Trunk D to take a country road over the bluff top to Highway 80

South, which would lead to Richland Center. As his old pickup labored up the steep hillside, he marveled at the dense maple, basswood, and oak forest blanketing the slopes. The magic of the Driftless was that these slopes, which are too steep for farming and of little benefit for pasturing dairy cows, had returned to thick second-growth forests. They created one of the softest and most pleasant yet rugged landscapes within the state, not to mention an excellent habitat for whitetail deer and wild turkey. He slowed the pickup to take in the last gasp of a spectacle that played out every spring in the densest of the forests in the area. From April through mid-May, the spring ephemerals suddenly force out their flowers before the thick forest canopy fully leafs out to shade and darken their habitat on the forest floor. In certain special places—this hillside being one of them—the trillium and Dutchman's breeches completely blanket the forest's understory to create a spectacular carpet of white blossoms. With Memorial Day approaching, the show was losing a bit of its luster, yet it still reminded Denny why he loved the environs of his home and farm.

Pulling into Richland Center, he parked the truck in the lot behind the bank and quickly stepped up and through the front doors, greeting the two tellers working the counter, friends he saw every Sunday at church. As Denny continued back toward Tom's office, he marveled at the stately interior of the old building, the polished marble tellers' windows and well-maintained woodwork, and the stark contrast it provided with the rest of the worn structures on Main Street. "No fairness in these times," he thought to himself. "The banks and CEOs are thriving while the rest of the valley is going down the toilet."

"Good to see you, Denny. Come on in and take a seat," Tom said as he jumped up from his desk to greet Denny with a firm handshake at his office door. Smartly dressed in a blue blazer, starched white shirt and tie, Tom looked the part of a successful banker and provided quite a contrast to Denny, who had put on clean blue jeans and a freshly pressed flannel shirt for the occasion. Tom was well known and admired in the community, the bank underwriting and supporting many of the philanthropic charitable activities occurring in the county. Although genuinely thought of as a friend of the farming community, the recent great recession had put some pretty severe constraints on the bank's lending abilities.

Starting with the obligatory small talk, Tom inquired about Mary and the kids. And with a note of true sincerity, he said, "I wish Mary had joined you so I could thank her for taking such good care of my mom and the other residents at the home. She's a godsend, Denny."

"Thanks, Tom. I will pass the message along to her tonight. She takes her work seriously at the home and thank God for the income, but I do worry about her, and that's part of the reason I'm here today. The poor woman gets up every morning with me at five-thirty to do the morning milking, moves on for the care and feeding of the heifers, and is into the car and off to the second shift at the home by two-thirty in the afternoon. She's been a great sport about it, but I'm concerned that the demands of the farm and the stressful job at the home are taking a toll," said Denny, hoping to lay the foundation for what he had to ask from his friend Tom. "You know as well as I that a farm like ours milking a hundred cows should have two individuals working full time, but I'm having trouble paying Sam for his help with the evening milking, much less being able to employ a full-time hand. I could probably increase my yields by at least ten percent by moving to three milkings a day like the big operations, but there's just no way I could manage that on my own."

Tom furrowed his brow with concern and said quietly, "Denny, I can only imagine the tight spot you are in. With milk prices being what they are, my hat's off to you for keeping things together as long as you have. We've both seen the farmers up and down the valley sell their herds, lease their land to the bigger operators to pay off their operating loans, and try to find work off the farm. You're one of the survivors, Denny."

"Not for long if you can't help me out, Tom. Quite honestly, I'm in arrears with my payment to the co-op, and Randy paid me a visit yesterday threatening to cut off feed shipments. In normal years, I could probably make it through the summer with minimal supplements and pasturing the herd, but after that God-awful flooding last year, I don't have enough corn and hay reserve to get me through to this year's harvest." Denny looked down at the floor, saying, "Is there any way you can see clear to float me some additional cash on my operating loan? They're saying milk prices have bottomed out, so if I could just get through the summer."

"Let's take a look at things, Denny," Tom said as he opened the thick file on his desk. "Oh, yeah, back in 2006 we issued a fifteen-year capital loan for $200,000 for your barn addition to increase the size of your herd from forty to a hundred animals. As I recall, with milk prices approaching twenty-four dollars per hundredweight, this move paid off nicely in the short run. But then, when the Feds decided they would add ethanol to gas, driving up the cost of corn, the feed bills went sky high. Then we came face to face with the recession of 2008, which hit everyone in the valley hard, including this bank. If memory serves me, Denny, there was almost a year where we allowed interest-only payments on your loan."

Denny thoughtfully nodded his head. "I remember those times well. Just when we thought things were going to get better, the industry was looking at over-supply problems. Lately, I'm lucky to see fifteen dollars per hundredweight. Heck, I need eighteen dollars to simply break even with the costs of production. Do you know what it's like to lose money every month? But they're saying there's light at the end of the tunnel. If I could just have an additional $50,000 on my operating loan, I think I can make it through the summer, Tom. I'd ask for a lot more and really go big with a new capital loan, but I barely have enough land to manage the manure from a hundred cows, so I don't see any way that I have that as an option on those steep hillsides along the West Pine."

"I don't know, Denny. If you can't cover the costs of your current liabilities, it's hard for me to support an increase to your operating loan. Lending is still tight after the recession, and, candidly, small dairy farms are not a good risk. On last week's afternoon ag report, I heard that Wisconsin lost more small farms in 2019 than at any time since the Great Depression. I'm afraid I'm starting to agree with the U.S. Ag secretary, who demoralized every small dairy farmer in the state at the recent Ag Expo by declaring that the future was to get big or get out. I'll bring your request to the committee, but I am going to be honest with you, I am not optimistic. How about if I give you a call on Thursday and let you know the verdict. But like I said, Denny, I am not hopeful."

"That's all I can ask, Tom. Please give me a fair shake. You know there's been a Schulz farming that land in Woodstock for the last four generations. It is starting to weigh pretty heavily on me that I may be the last." Shaking hands across the desk, Denny got up to leave. Pausing in the doorway for a minute, he asked, "Tom, one other thing. Could you please make that call after two-thirty in the afternoon? I'm trying not to worry Mary about this, so if you could call after she leaves for work, that would be helpful. Thanks, and please give my best to your family."

Putting Off That Talk

Denny tossed and turned in bed in a sleepless turmoil of worry after another evening of too many Jack and Cokes. A reoccurring pattern of sleepless nights worsened by increased alcohol consumption was taking a toll on his psyche. Lying awake on his back, he saw the headlights from the little Chevy reflected off the ceiling as the car bounced down the lane, returning Mary from her shift at the nursing home in Center. He listened quietly as Mary entered the back door and dropped her keys on the small table in the back entryway. Reaching the top of the stairs, she turned to enter the bathroom. After she took a couple of minutes to undress, Denny heard her turn on the shower and step in. With the experience of more than thirty years of marriage, he knew this as a sure sign that his beautiful, sweet-scented wife would crawl into bed, interested in a little loving after a long, hard shift at the nursing home. "Should this be the moment of truth?" wondered Denny. As much as he craved her soft embrace and the desire to share the complete story of their money problems with her, he desperately wanted to wait for an answer from the bank, hoping beyond hope that the story might have a good ending. Feigning sleep, he turned away from her side of the bed as she quietly slipped in beside him. She gently rubbed his back, hoping to wake him from his slumber, but Denny resisted the urge to turn over and draw her close to him. No, the talk would have to wait for some workable options. He didn't want to burden her with a bunch of worries that he had yet to resolve. Eventually, Mary gave up, turned over, and he soon heard her fall into a deep sleep, marked by a slight snore. Denny was not so lucky. Sleep eluded him as the tapes of a disastrous outcome for the farm played over and over in his mind.

Pete Hammes

Pete bounded up the steps of the old county courthouse, anxious to get the evening meeting over with. A well-dressed, vigorous, fit forty-year-old who had translated a degree in Ag Economics into a successful large dairying operation, he had to explain to the community in a public meeting this evening his plans to expand his family's operation, the Broad Valley Dairy, from around 700 Holstein dairy cows to about 2,000 animals. Richland County had joined other counties in the state in passing local ordinances around the siting of large dairy operations, called, in the vernacular, confined animal feeding operations or CAFOs.

Reaching the top of the stairs, he moved quickly down the hall to join his county board rep, Harvey Johnston, who would be moderating tonight's meeting. "Evenin', Pete," Harvey greeted. "Are you ready for this? I am not anticipating a pleasant meeting."

"Yeah, Harvey, I am afraid this is going to be painful, but I really appreciate your willingness to help keep things under control this evening," Pete exclaimed.

"If this goes like other siting meetings around the state, we can probably expect PETA and the animal rights activists, the groundwater protection advocates, as well as some particularly cranky neighbors making all kinds of allegations about the inhumane treatment of your herd, well pollution, and odor problems, some of it likely to be very heated. Do you have anyone lined up to speak in support of your plan?" Harvey asked.

"I sure do, Harvey. George, my vet, will speak to the care of my herd; Sarah, our local DNR water specialist, can talk about my past compliance with my permit conditions as well as future restrictions that will be part of my expanded permit. The State Department of Ag advocate will talk more generally about large dairying operations' safety and economic benefits. Manny Hernandez is prepared to represent my employees and talk about the additional jobs that the farm will support," explained Pete. "Also, I will talk about future plans we are working on to try and incorporate a manure digester into our operations, which should help with concerns about land spreading, water, and air pollution. Finally, Bob from the electric co-op can explain to folks about the benefits to the community from the power generated by the digester that will potentially be fed back into the electrical grid."

"That sounds like good planning," said Harvey as he noticed the large numbers of folks filing into the hearing room. "Pete, you are a well-respected and important part of our community. We'll get through this tonight. Just remember, if you are following your permit conditions, there is not a hell of a lot that will result from this hearing. It will likely be emotional, and a lot of allegations will be thrown around, but in the end, this is a Right-to-Farm state, and our state laws protect operations like yours. We better get moving. I think it's show time."

The Lower Pine River Valley

The West Branch of the Pine River carves beautiful cliffs and outcroppings and flows beneath a large bluff in the village of Rockbridge, forming one of the most unusual and impressive geological sites in Wisconsin: a natural bridge made from massive sandstone rock. After passing underneath the bluff, the river broadens as it joins the Eastern Branch and slows to approach and then passes through the county seat of Richland Center. The cold-water fish found in the river's northern reaches are replaced by warm water-loving species like sunfish, bass, and carp. The valley widens considerably south of town as it meets the vast sand plain surrounding the Wisconsin River Valley. It is in this part of the county that you find the larger dairying operations.

Research from the University of Wisconsin School of Ag indicates that it takes one to two acres of land to manage and spread the manure from one dairy cow responsibly. Dairy cows produce slightly more than one hundred pounds of waste a day. Thus, the smaller farms and steeper slopes of the Upper Pine River Valley are not well suited to the trend toward larger and larger dairy operations. Spreading cow manure on the steep bluff sides of the upper valley allows nutrient runoff to pollute the streams and river. However, south of town in the broad valley, the Hammes farm family and others have been able to acquire neighboring farms and properties and assemble the huge, flat tracts of land that allow for raising enough feed and corn silage and enough acreage to spread and manage the manure from larger and larger dairy herds.

South of town, the few remnants of the iconic red Wisconsin dairy barns are crumbling, having been relegated to hay and equipment storage structures and replaced by very large, single-story, sprawling free-stall industrial-looking barns. The cows are no longer constrained in stanchions during milking but can move about between areas providing ample, clean bedding and a feeding line. Often the cows are no longer pastured in summers on grasses outdoors, spending their entire time on the farm in the free-stall barns. Reportedly, this type of barn is more comfortable for the cows than the small, cramped stanchions of the old dairy barn. Temperatures are kept warmer in the winters through heat generated by the cows, and large fans and evaporative coolers keep the cows more comfortable in the summer heat. Manure is automatically removed and conveyed to the manure pits, where it awaits being spread on the land before spring planting and after the fall harvest. The cows' health and lactation status are monitored via electronic devices akin to a Fitbit, fixed to a device looped around the cow's neck. Veterinary care, such as insemination of the animals, calving, trimming of hooves, and attending to sick animals, is more easily provided in the larger structures. All these precautions contribute to the efficiency of the operation and the avoidance, as much as possible, of the backbreaking work associated with dairying operations conducted in the smaller dairy barns.

But the greatest advance, and by far the most expensive component of the large dairying operation, is the sophistication of the milking parlor. Instead of the dairyman having to stoop to lower the pneumatic milking machine and attach it to the cows' udders while secured in the stanchion, the cows are herded three times a day to the elevated platform of the milking parlor. A team of two to three milkers no longer have to bend down to wash and sanitize the cows' udders. Instead, they work from the parlor floor at chest height to the cows and attach the milking machine, often conveying the milk to a refrigerated five-thousand-gallon tank trailer. Operations with larger herds may milk around the clock, requiring three shifts of parlor workers. The large operations no longer need dairies to drive a tank truck from farm to farm to pick up the milk from small, refrigerated milk storage tanks. Now, when full, a large semi will be dispatched to the farm to transport the tanker of fluid milk directly to the dairy or cheese plant, saving in transportation costs. Even greater levels of sophistication have been introduced into some very large farms, such as robotic milking machines that reduce the need for hired help in the milking parlor. Also, new electronic monitoring technology can record the volume, milk-fat yield, and protein content of the milk produced by each animal in the dairy herd while in the milking parlor.

The large operations can increase milk production by optimizing the genetics of their herds. In the early days of Douglas and Henrietta Schulz's farm, a cow produced about three thousand pounds of milk per year. Today that production has increased to about twenty thousand pounds. This increase, together with improved animal care and careful feeding and nutrient control, greatly reduces operational costs to allow the larger farms to compete successfully in what is viewed as an increasingly cut-throat commodity market that is squeezing the small operator out of business. Farm economists predict that this is the future of the dairy industry in Wisconsin.

The sheer size and stark contrast to Wisconsin's historical dairy traditions provide fodder for what many see as the very troubling movement of the industry toward confined animal feeding operations or CAFOs. Legitimate concerns exist, such as manure runoff from land spreading, degrading surface waters, streams, and rivers. Excessive manure land application on the thin, sandy soils of the lower Pine and Wisconsin River valleys can lead to groundwater and drinking water well contamination. And one must only drive past the large dairies on a warm, humid summer evening when the barn fans are in full operation to appreciate neighbors' real concerns regarding air quality and odor control. Finally, many feel that large dairy operations create an environment that is inhumane to the animals, especially given that the cows spend the majority, if not all, of their time at the farm confined to the free-stall barn.

These concerns were passionately expressed at the public hearing held around the proposed expansion of the Broad Valley Dairy operation.

The milking parlor at Mystic Valley Dairy provides an efficient and sanitary around-the-clock process for managing the milking of a larger dairy herd. Copyright 2020 John P. Gauder.

The free-stall structure is rapidly replacing the iconic early Wisconsin dairy barns. Taken at Mystic Valley Dairy. Copyright John P. Gauder, 2020.

Millions of years of erosion from the East and West branches of the Pine River have carved the magnificent sandstone cliffs found in the upper river valley. Copyright 2020 John P. Gauder.

Passions Heat Up

As temperatures and emotions in the stuffy hearing room heated up, it occurred to Pete that his hope of introducing friendly voices to counter those opposed to the expansion of his operations was proving a bit futile. He leaned over to Harvey to quietly ask, “How much more of the pain will I have to go through?”

“The good news,” Harvey whispered, “is that we only have two people remaining who registered to speak. The bad news is that one of them is the state spokesperson for PETA.”

Pete shifted uncomfortably in his chair, steeling himself for the worst. He was thinking, “Thank God, the final speaker this evening will be my employee Manny, who hopefully can end things this evening on a high note.”

Harvey stepped up to the microphone and announced, “The next registered speaker is Marcia Dombrowski. Marcia, would you please step up to the microphone, introduce yourself and keep your comments to ten minutes or less.”

The crowd, which was already starting to thin out, moved about uncomfortably in the hard wooden chairs as a well-coiffed young lady approached the microphone.

"Good evening. My name is Marcia Dombrowski from Milwaukee, and I am the state spokesperson for PETA, an organization dedicated to the protection and ethical treatment of animals," the young woman confidently read from a prepared statement, obviously comfortable speaking in front of a crowd. "Our organization, representing the voices of over 10,000 people within this state, is very much opposed to the proposed expansion of Pete Hammes' operation, Broad Valley Dairy. While I am speaking tonight in opposition to issuing a revised permit for Mr. Hammes' proposed expansion, PETA as an organization is opposed to the dairy industry in general."

The men and women in the audience sat up and took skeptical notice of Marcia's comments. While several of the speakers had expressed concern about the potential for air and water pollution stemming from a larger operation, this was farming country, after all. As troubling as the times were, farms and the dairy industry still provided the community's economic foundation. Audible groans were heard from the crowd, followed by multiple, disbelieving eye rolls.

Marcia continued to read her statement. "And the cows are artificially inseminated each year for the three to four years of their milk production. Calves are separated from their mothers immediately at birth. Male calves have no future in the dairy industry, so they are often sold to the veal industry, where they are kept under inhumane conditions and slaughtered while still young. Female calves are removed to heifer rearing facilities, away from their mothers, until they can be inseminated and brought into milk production. All the while, cows are kept within the free-stall barn in crowded conditions and then, after about four years of milk production, are sold to slaughterhouses for meat ..."

Pete had heard it all before and was zoning out as he went over all the reasons and facts that he could use to rebut these tiresome allegations. As is the case in every industry, he knew there were a few bad operators who probably deserved this kind of criticism. But, dammit, she and her PETA buddies simply did not understand the dairy business. A good dairyman had no reason not to provide the best care for his animals, especially if he hoped to run a profitable operation. Carefully monitored herd health and genetics, dedicated veterinary services, the best and most nutritious of feeds as well as comfortable lodging in the free-stall barn characterized his operations. But given the fervor of the present speaker, he knew it was futile to try and present a counter-argument. As an old Irish friend had once told him, "You might as well save your breath to cool your soup."

Marcia completed her statement by saying in a louder, firm voice, "For these and other reasons related to the inhumane treatment of dairy cows, PETA wishes to go on record in opposition to the expansion of Broad Valley Dairy. We urge the Wisconsin Department of Natural Resources to reject this revised permit application. Thank you." She returned to her seat.

Little applause followed these statements as the crowd obviously was becoming antsy and was generally disapproving of the last speaker. Harvey quickly rose to the microphone to announce the next speaker of the evening. "The final speaker who registered this evening is Manny Hernandez. Mr. Hernandez, would you please step up to the microphone, introduce yourself, and kindly keep your remarks to ten minutes or less."

A compact, muscular man in his mid-thirties, wearing new blue jeans and a crisp, white, ironed shirt approached the microphone. He hesitated a bit, clearly uncomfortable in addressing the crowd. "Good evening. My name is Manny Hernandez, and I am an employee of the Broad Valley Dairy," he said in a lightly accented Hispanic voice, as he also read from a ragged piece of paper. "I have been an employee of Mr. Hammes for over nine years. I came to this country from Chiapas State, Mexico, twenty-seven years ago with my parents when I was three years old. I'm in the U.S. legally and will be eligible for full citizenship after the next several years. After graduating from high school, I moved to Richland Center to take a job in Broad Valley Dairy and registered as a Dreamer. Pete has provided me and the other five men who work the milking parlor with good jobs, paying a good wage, and offering reasonable benefits to support our families. He provides housing to several of our families. A year ago, Pete agreed to sponsor me for permanent residence in the U.S., and I received my green card, something very important for my family and me and for which I owe him a debt of gratitude. We run a very good operation, and we are proud of this dairy. If he can expand, I believe he will need to add at least two additional employees for each of the three shifts in the expanded milking parlor. More good jobs for our area."

Manny hesitated as he looked up at the crowd. Clearly wanting to wrap things up, he completed his short statement by saying, "I urge you to support the dairy expansion so we can bring additional jobs and economic benefits to our community. Thank you."

His statement was by far the briefest made this evening, but at least it was supportive. As Manny turned to go back to his seat, Pete caught his eye with a nod and a mouthed "Thanks" as he breathed a sigh of relief that the ordeal was over.

Harvey quickly jumped to the microphone and concluded by saying, "I'd like to thank you all for turning out this evening and those of you who spoke. I recognize that this is an important issue for our community and many of you. The county board will next review the proposal, the record from tonight's meeting, and the management plan submitted by Pete Hammes. We will make a recommendation to the DNR, the ultimate decision-making authority, as to whether to grant the permit. So, stay tuned and, again, thank you and good night."

Pete stiffly got out of his chair and went over to shake Harvey's hand. "Thanks for moderating this sideshow," he said sarcastically. "I'm just glad it's over with."

After many years on the county board, Harvey had hardened to the theatrics of public meetings. "Just make sure your management plan and operations are in compliance with the CAFO regulations, Pete," he counseled. "And as I said earlier, our Right to Farm statute gives you the edge you'll need for your new permit and expansion. Besides, new jobs and expanded operations are what this county needs during these shitty economic times. Hang in there, and we'll be in touch," he said quietly as he quickly headed for the door, smiling and shaking as many of his constituents' hands as possible as he left.

Before Pete could turn to leave, he noticed a familiar face approaching him with a stooped characteristic of an older dairy farmer. He recognized Denny Schulz, who ran a small dairy operation up the Woodstock way. "Hello, Denny. Please don't tell me that you came this evening to try and put the kibosh on my proposed expansion?"

"No, Pete. I was just curious as to how things would go and wondering if your expansion might help me out."

"How so, Denny?" Pete curiously asked.

"Well, Pete, you know how tough the past couple of years have been to us dairymen. And my small operation doesn't have the luxury of the type of expansion that was talked about here tonight. Not that I would want to be put through the wringer like you were," said Denny, giving Pete a sympathetic glance.

Denny continued, "I can't begin to meet my operating expenses. I am so far underwater that I don't see any way out. My Hail Mary pass was a visit yesterday to see Tom at the bank for additional credit on my operating loan. But he wasn't hopeful, so I'm expecting bad news on that front soon." Denny peered down to avoid looking Pete directly in the eye, obviously uncomfortable about what he was to say next.

"Pete, I can't believe I am even saying this, but I think the only option left for Mary and me is to sell the herd, lease our land, and get the hell out of the business," Denny blurted out.

Pete could see the anger and hurt in Denny's eyes. As a fellow dairyman, he knew that this option was truly the most painful decision Denny could make. Pete gently put a hand on Denny's shoulder and let him continue.

"If I get the bad news from Tom, which I'm fully expecting, I was hoping you might be in the market for my Holstein herd and heifers. It's a good herd, and you know that I am very careful of their genetics. I can't bear the thought of sending the girls up to Green Bay to the hamburger factory. I hate to say it, Pete, but I'm getting desperate," confessed Denny.

Denny's plight erased the discomfort that Pete had experienced during the evening hearing. "Denny, I've known you a long time and know you run a good operation. You and Mary are good members of our community, and our kids have grown up together. If this permit expansion goes through—and I expect that it will—you need to know that I will do anything to help you."

Denny took a deep breath. "Thanks, Pete. That's a relief. I'll be praying like hell that I don't need to call on you. But I will let you know what I hear from Tom." With that, Denny looked Pete in the eye, extended a sincere, firm handshake, and quickly turned to go.

"Hang in there, Denny," Pete called out. "The whole county is pulling for you."

Manny Hernandez

Manny quietly greeted some friends and neighbors as he found his way out of the old courthouse hearing room and walked across the street to the Mexican restaurant which had recently opened on the courthouse square. Opening the door, he immediately relaxed, recognizing familiar faces among the bar and waitstaff. "Hola, Manny. ¿Qué pasa?" greeted Maria, a cute, young Latina from behind the bar. As Manny approached, he explained, "I just finished talking at the hearing about the Hammes farm in the courthouse, and I could really use a beer!"

"I was wondering what was going on over there, with all the cars parked downtown," said Maria as she expertly topped off the Especiál draft with a nice head. As she slid his glass across the bar, she asked, "How did things go?"

"Most folks are not happy about it, but I'm sure glad it's over with. Lots of people complained about manure spreading, farm smells, cruelty to the cows, and things like that, but Pete is pretty sure that things are going to work out," explained Manny.

"Good," said Maria. "A lot of our compadres, including you, Manny, depend on that dairy for jobs. José has a couple of younger brothers who would like to come north if the dairy needs to bring on more hired help."

"Ya, but this is not the time to try and make that border crossing," cautioned Manny. "With all of the focus on illegal immigration, everyone is saying that crossing now is darn near impossible."

"If we look at all of the presidential election signs up and down the valley, it makes you wonder if most of the farmers support those policies," wondered Maria.

"I just don't understand it. I know that most dairy farmers support some type of immigration reform. Still, their political party seems to direct a lot of hate at our amigos and immigrant labor," said Manny, with a quizzical look on his face. "Most dairy operators truly appreciate our work, but heck, a few big ICE raids and tighter border controls could completely shut down the whole dairy industry. Without us, there would be nothing left in this valley but idle dairy barns."

"You're right, Manny," said Maria sadly. "I would love to go back to Chiapas to visit my madre and my abuela, but God forbid I go south and then can't get back over the border. If it weren't for FaceTime and Skype, I would never see them! Do you want another draft?"

"No. I better be heading home. I'm sure by now that the kids are in bed, but Kathy is up waiting for me," said Manny, as he put a five-spot on the bar along with an extra buck for Maria. "Thanks, Maria. How about if I bring the family in for lunch Sunday after church?"

"That would be nice. Adiós, and give my best to Kathy," Maria said as she wiped down the bar.

As he got up from his barstool, Manny quickly greeted other friends and neighbors sitting in booths along the wall. "This restaurant is quickly becoming the go-to place in town," he thought as he headed out the door.

The Hired Hand

There was a day when the family members, with a combination of grit and hard work, could manage the workload of a fifty-cow dairy farm. However, as operations and the herd sizes grew in the 1990s, so did the need for additional hired hands on the farm. To hold down operating costs to the greatest extent possible to run a profitable operation, farmers turned to immigrant workers to meet their labor needs. Also, because as they searched for help, most farmers found that few, if any, U.S.-born workers were willing to put up with the low wages, long hours, hard work, and difficult conditions found on the farm. Some have estimated that immigrant labor makes up forty to fifty percent of the employees within the dairy industry. However, many farmers suggest that these estimates are low.

It is tacitly acknowledged that many of these laborers are undocumented. They make the dangerous journey of more than two thousand miles north, including a perilous border crossing, to escape poverty and hardship in their native lands. Starting wages of ten to twelve dollars per hour at a farm in Wisconsin far exceed the wages they can realize south of the border and even allow some to salt away money to send home to help their families.

In Manny Hernandez's case, he and his two sisters were Dreamers (undocumented children) of less than five years old when they came with their parents on the perilous journey north from the southern reaches of Mexico in 1990. The Hernandezes found a welcoming family of Mexican immigrants in south Chicago and plenty of opportunities for unskilled, hourly work in the nearby mills and industries. The decade 1990 to 2000 saw a sixty-nine percent increase in Mexican Americans in Chicago, resulting in Chicago having a greater population of Mexican Americans than Houston or San Antonio, Texas.

After graduating high school in Chicago, Manny heard about the availability of jobs in Wisconsin's dairy industry through friends and neighbors who had moved north and found himself in Richland Center working at Broad Valley Dairy. Initially, he lived with three other immigrant Broad Valley employees in a somewhat dilapidated and cramped mobile home that Pete Hammes had moved onto his land to house his hired help.

Dairy operations are marked by long hours, backbreaking work, and dangerous work conditions. The U.S. Occupational Safety and Health Administration (OSHA) has no jurisdiction over small family farms, so safety statistics for this industry are fragmented and incomplete. Some studies, however, put the dairy industry near the top of the deadliest U.S. occupations. Manny, however, did not mind the hard and often dangerous work. He enjoyed caring for the cows, most of whom were gentle, docile creatures. Most importantly, he was thankful that he was able to earn a good wage away from the crime and gang problems that had crept insidiously into his former south Chicago neighborhood. Rather quickly, he found himself embraced by the growing Hispanic population that now supported its own restaurant, grocery store, and church in the Richland Center area.

Manny met Kathy, a local girl, at the county fair three years after moving north to work in the dairy. After a whirlwind courtship, they married in Manny's church in front of many Hispanic friends and neighbors and a very skeptical family of the bride who was more than a little disapproving of this union with an "illegal alien." Pete Hammes provided the newlyweds with housing in a small home that came with an adjacent farm he had purchased to expand his operations.

The early years of the marriage proved difficult, requiring that the minimum wages provided by the dairy be supplemented by Kathy's full-time job. Although Manny registered as a Dreamer in 2013, the family lived in constant fear that any speeding ticket or similar small crime or indiscretion could threaten his continuing to live in the U.S. With the arrival of two children and a U.S. congress unable to agree on a path to citizenship for Dreamers, Manny and Kathy faced the daunting challenge of expensive legal bills necessary to navigate the archaic and circuitous process required to gain permanent residence. Manny was eligible for a green card because he was married to a U.S. citizen and had an employer sponsor in Pete Hammes. But the process was unnerving and tedious and required the very expensive assistance of a Milwaukee-based immigration attorney.

After several years of endless waiting, Manny's attorney called to inform him that all was in order for him to take his final steps toward a green card. However, Manny's initial jubilation was immediately offset by the process he would be required to go through. He would have to return to Mexico and apply through a U.S. Consulate in his native country. While the attorney assured Manny that the process should go quickly, he worried about how long he would have to remain in Mexico away from his family, given his experience with the slow-moving immigration system. And, of course, in the back of his mind was the main concern that a system seemingly hostile to immigrants would find some reason not to approve his application, and he would be unable to ever return to the U.S. to be with his family.

On June 10, 2019, he received an email from the U.S. Consulate in Ciudad Juárez immediately over the U.S.–Mexico border from El Paso, giving him a date and time two months in the future to report to the embassy for his application for a green card. After receiving repeated reassurance from his attorney that it was now or never, Manny checked his paperwork for the last time, including a detailed immunization record, kissed his wife and hugged his children before getting into his old Dodge pickup for the long drive to Juárez. After submitting his application to the U.S. Consulate at the appointed time and two weeks of an agonizing wait in a cheap motel, miracle of miracles, the ponderous system moved to approve his application. On August 25, Manny headed north and made the crossing back over the border, this time legally with his new green card.

The Accident

Manny clocked into his morning shift at 6:50 a.m. and high-fived the two fellow Latino workers eager to complete their third shift in the milking parlor. "Any equipment or other problems that we should be aware of, amigos?" he asked them as they entered the break room. One of the young hands, Miquel, looked contemptuously over at his third-shift colleague and scoffed, "Carlos showed up late again last night, so we fell behind in the parlor. You are going to have to bring in the last of our shift's cows before starting your own. Sorry, amigo."

Each of the three shifts had to run the entire herd of about seven hundred cows through the milking parlor, making for an extremely busy eight-hour workday. It wasn't uncommon to have to put in extra hours due to additional tasks, such as feeding and manure clean-up operations. Typically, the last of the cows brought into the parlor were the younger animals, often not yet used to the milking routine and sometimes a bit more difficult to handle. Manny cursed softly to himself, knowing that this would make for a busy work shift.

"And keep your eyes on Diabla," warned Carlos. "She's become more skittish than ever and still is hard to get into the parlor stall." Diabla, or she-devil, was a nickname all of the hands used to refer to a new Holstein that had just calved and was brought into lactation for milking. While most cows docilely adjusted to the milking routine, it seemed that with each new group of cows rotated in, there were always a few that had a harder time adjusting. At about thirteen hundred pounds, an ornery Holstein was a force to be reckoned with, and Diabla lived up to her nickname. Quick to kick at anything or anyone approaching her from the rear and resistant to moving into the milking stall, she was proving difficult to handle. "Will do, Carlos. Thanks for the warning, amigo," said Manny as he quickly pulled on his coveralls and stepped into his barn boots.

To get a jump on his work before his first shift co-worker arrived, Manny headed into the free-stall barn and immediately saw the couple of dozen animals from the third shift yet to be brought into the milking parlor. Manny walked to the front of the group, keeping a wary eye on Diabla as he stepped beside her to open the metal gate to the milking parlor. As is not uncommon among the younger cows, Diabla suddenly jumped onto the back of the cow in front of her, trying to mount her. Manny cursed loudly at her in an attempt to settle her down. The unreceptive cow kicked and reared violently, throwing Diabla against Manny and pinning him against the fence. The last thing Manny remembered was the vision of the black and white creature coming down on top of him and the evil look in the Holstein's eyes. Then everything went dark.

The other first shift hand, sixty-five-year-old foreman George Harriman, had sold his herd, leased his land, and gotten out of the dairy business eight years earlier. More than a bit bow-legged and stooped from years of hard labor in his old dairy barn, he nonetheless enjoyed the business and was happy to continue to work as hired help in Pete's dairy. He pulled up to the barn just as the third shift boys were pulling away in their Toyotas, changed into his barn clothes, and set about looking for Manny to start their shift. Not seeing Manny in the parlor, he headed out into the barn to greet his co-worker and begin their work shift. George liked working with the young Hispanic fellow. Manny was a hard worker, very dependable, and more than carried his weight with the heavy, quick pace of work required to move the herd through the milking parlor.

George found it odd when he didn't see Manny in the barn. It was unlike him to be late for his work shift. If there was a problem with his young family or at home, which was unusual, he could always count on a call well before the shift started. Sensing some unusual agitation among the cows waiting to be brought into the parlor, George headed over to herd them up the ramp onto the milking platform. It was then that he saw the horrifying cause of the animals' unusual behavior. Lying face down in the manure was the body of Manny. Hurrying over to him as fast his arthritic legs could carry him, George quickly turned Manny over and was immediately relieved to find him breathing but unconscious. "Can you hear me, buddy?" he anxiously called out. Not hearing any response, the burly farmhand quickly but gently lifted Manny over his shoulder and carefully moved him away from the animals to the center aisle of the barn. He pulled his cellphone out of his pocket and first called 911 for an ambulance. He then quickly dialed Pete to let him know of the accident.

"How bad is it, George?" Pete asked.

"There's not much blood loss, his breathing seems strong, but he is not conscious," answered George. "I'm guessing he got crushed pretty bad between a cow and the metal gate."

"Let's not move him until the paramedics arrive. Do you think the Center ER can handle it, or should we have him transported to Madison?" asked Pete, who worried about the limitations of a small, rural hospital.

"I suspect we ought to let that be the call of our local docs, Pete," answered George as he heard the far-off siren of the ambulance starting to head south on Highway 14 toward the dairy.

"Do you want me to ride the ambulance in with Manny?" asked George.

"I'll be right down to the barn, and I'll go in with him, George," answered Pete. "I've got all the insurance info, and I'll call his wife, Kathy, on the way. Besides, those cows gotta be milked. I'll give the other hands a call to see if someone can come in to help you. Keep your fingers crossed that all turns out okay. I'll let you know how things stand as soon as I get some information. Oh, and one other thing. Do you suspect that it was that young Holstein that the hands are calling Diabla?"

"I suspect so," surmised George. "She was in the area waiting to be moved into the parlor where I found him. And, Pete, as we have been saying, she is a nervous bitch."

"Say no more, George. She is on her way to the hamburger

factory. I'll be in touch," signed off Pete. ❧

Worry Like No Other

Manny's wife, Kathy, was at home going through the morning routine of getting their two young children dressed and fed. She was playing whimsical word games with them, determined to bring them up in a bilingual household. The call came in as she got the children packed and ready for childcare at her mother's house.

"Kathy, this is Pete over at the dairy. I've got some bad news. Manny was in an accident this morning and is being taken over to the Richland Hospital ER," Pete cautiously explained.

"Oh, my god, Pete," Kathy cried out. "What happened? Is he going to be all right?" she asked, toning down her voice so as not to alarm the children.

"Manny was alone, so we aren't sure what exactly happened." Trying to reassure her, he said, "I'm sure he'll be fine. We're pulling into the ER now so I can meet you here." He didn't want to give Kathy false hopes, but the paramedics had shared with Pete that Manny's vitals were strong, although he still had not regained consciousness.

The two children were a bit confused over the emotional hugs and tears as they were hurriedly dropped off at their abuela's house. "I'll let you know what's going on as soon as I get there, Mom. Say a prayer for Manny," called out Kathy as she hurriedly got back into her car for the long drive to the hospital.

From a devoutly religious family, Kathy said a quick prayer for her dear Manny as she wiped tears from her eyes while driving. This confirmed all of her worst fears: something she had been warning Manny about over the last several months. He had to get out of that job! The backbreaking work, the stories of accidents and near misses at other dairies in the valley that circulated among the Hispanic community, coming home filthy and smelling of manure. This was not the future she envisioned for Manny or her family. But Manny never complained. He was dedicated to that dairy to a fault and was ever so thankful to Pete Hammes for providing him a job and sponsoring him for his green card. No, her sweet husband, despite her protestations, had no intention of leaving the dairy.

As she hurried through the self-opening ER doors, she immediately spotted Pete. He was talking in hushed tones to the handsome young doctor whom she recognized from a picture featured in the local shopper announcing that he was opening a practice in the community.

As she walked up to them, Pete immediately embraced her with a warm hug, whispering in her ear, "Everything is going to be all right." While holding on to her hand, he said, "Kathy, this is Dr. Aryan Patel. He is on duty today in the ER and is providing Manny excellent care. Doctor, this is

Manny's wife, Kathy."

Kathy held out her hand to Dr. Patel, which he eagerly shook. "It is very nice to meet you, Kathy. Please call me Ari," he said in a very reassuring voice.

"How's Manny doing? When will I be able to see him?" asked Kathy, with eyes that were red with worry.

"I am pleased to be able to report that Manny is doing well. He is going to be a very sore young man for several weeks as he has multiple broken ribs on his left side. But the ribs did as they were designed to do; they protected his vital organs from blunt trauma. Although we suspect there is some bruising, we do not see significant internal bleeding associated with severe damage to the lungs, liver, or spleen. He also took a nasty blow to the head, which Pete figures may have been caused by the metal gate as he fell backward. He has regained consciousness, but we will be keeping him under observation for a while to see if there might be repercussions from the head injury. He is a lucky young man, Kathy, and we expect him to make a full recovery."

With sobs of relief, Kathy thanked the young doctor and again asked to see Manny. "We are finishing up our assessment and then want to clean him up to be presentable for his wonderful young wife," teased Dr. Patel in an attempt to bring some levity into a delicate situation. "So please take a seat, and I'm sure you will be able to see him momentarily."

As the doctor turned to get back to his patient, Kathy tearfully thanked him again. Pete came over and gave her another consoling hug, saying, "Rest assured, Kathy, that our insurance will cover the medical bills, and Manny can take whatever time is necessary to recover fully." And then he added, "I think the Center Hospital is damn lucky to have such a competent new doctor."

Kathy agreed, thanked Pete, and moved to a chair so she could call her mother and share the news of Manny's condition. While she was appreciative of Pete's support and concern, in the back of her mind, she was thinking that at long last, this might be enough for Manny to consider leaving the dairy.

The Moment of Reckoning

Denny awoke at his usual four-thirty on the Friday morning before Memorial Day. He felt anything but rested after a sleepless night, but the morning chores—most importantly milking the herd—was not something that could be put off. As usual, the cows already had returned from pasture and were waiting patiently in the barnyard to be let in for the milking. As he swung open the door to the barn, the cows slowly filed into the barn magically into their stanchions as if they were assigned seats in a school classroom. Denny gently caressed the flanks of several of his favorites and enjoyed the soft looks of their dark bovine eyes as well as the occasional lick he received from sandpaper-like tongues as they sauntered by him.

As he walked down the center aisle between the two rows of stanchions, he stopped to marvel at the beauty of the old barn. The original structure was built on a thick foundation of rock quarried from a nearby hillside. The whitewashed stone so expertly pieced together and mortared in place spoke to the skills of the original craftsman who built the barn, most likely friends and relatives of his great-great-grandfather. He never tired of admiring the rough-cut oak beams and columns that supported the haymow above, timbers that had been harvested from the native oak forests of the nearby bluffs and cut to size in the water-powered sawmill located along the West Pine. To complete the sensory experience, the soft lowing of the cows together with the mixed scents of sweet cow manure and hay from the mow above combined to remind him why he loved his solitary, early morning time alone in the barn. "This," he thought, "is why I need to keep this place afloat."

The division of labor on the Schulz farm was such that Denny would rise first in the morning, let the cows into the barn, and set out the feed and nutritional supplements for the cows. Next, he got the pneumatic milking equipment ready for the morning chores before Mary joined him in the barn, typically around five to help with the morning milking. The whole procedure was repeated around five in the early evening, but since Mary had taken the job in town, Denny had hired Sam, a retired dairy farmer, to come in each evening to help with the milking.

As was typical, Mary skipped into the barn with a smile on her face and a welcoming good morning kiss. Denny marveled at how she could maintain such a great attitude despite the hard work and limited sleep allowed by her second shift at the nursing home.

As they both gathered up the milking equipment and moved to the far side of the barn across the center aisle from each other, the interrogation that Denny dreaded began.

"Denny, you have been awfully quiet lately. I am starting to worry about you," said Mary as she quickly attached the milking machine to the first Holstein's teats. "I tried to wake you last night when I got home, hoping we could talk, but you were sleeping pretty soundly."

Denny admired the fact that Mary moved effortlessly between the cows, caressing their flanks with love and expertly moving the pneumatic machines down the line of cows in their well-rehearsed dance that was repeated each morning.

"Sorry, sweetie. I guess I was just beat after a tough day," Denny fibbed. Hoping to change the subject, he asked, "How are things at the home?"

"No news there, Den. So, what were you up to yesterday?" inquired Mary, seemingly eager to find out what was troubling her husband. Normally, they conducted their early morning milking chores in relative silence to the sounds of local country music radio station emanating from the old Zenith perched in a window near the center of the barn. Mary briskly walked up to the radio and snapped it off, signaling that the conversation to follow would be serious business.

Denny took in a deep breath and was getting a bit tight in the throat. He would have much preferred to have this discussion over the dinner table or on the front porch swing with a couple of good stiff cocktails but realized that there was no backing out now. The pressure of keeping their dire financial situation from Mary was taking its toll on him. Since their kids moved out of the house, they had grown closer and were complete partners in their farming business—

but, more importantly, also in life. It was killing him to keep secrets from Mary. "Oh, hell," he thought. "Maybe it's time to get it all off my chest."

Stepping between the next two cows in the stanchions, he quickly sanitized the old girl's udder and attached the milking machine before backing into the center aisle and approaching Mary, who was methodically moving the milking equipment down the line of the cows in the stanchions on her side of the barn.

Coming up behind her, he gently touched her shoulder as she was bent over preparing to milk the next cow. "Honey, you've known that our dairy business is tight, but I haven't been completely honest with you about how bad things are," Denny started. As Mary turned to look him in the eye, he simply blurted out, "I think we are going to lose the farm."

Mary was a strong woman, having grown up on a farm herself and very used to the tough life of a dairy farmer's wife. Things were always touch and go with regard to the farm's finances and lately seemed worse than ever. But with sheer grit and hard work, they had weathered the lows and vagaries of often ridiculous returns on their fluid milk production, taking comfort in building a life for their tightly knit family in the glorious environs that the Schulzes had farmed for well over a century. This farm was the anchor of their family and the anchor for the four generations of Schulzes that farmed it before them. With the loss of their dairy, she knew that Denny would be untethered from the way of life he loved. Mary also knew that Denny was troubled because neither their son nor daughter was interested in taking over the farming operations, but she now saw a worry in his eyes that shook her to her very core. "C'mon,

Den, we've hit hard patches before but always made our way through. Heck, that's why I took the job at the home," said Mary as she turned to grab Denny in a warm hug.

Denny eagerly fell into her embrace, sensing immediate relief from the pressure of keeping these secrets for so long. But, damn, he hated the encroaching feelings of failure, blaming himself that Mary had to maintain an exhausting job outside the home in addition to her numerous daily farm chores.

As Mary backed off from their embrace, she appeared shaken to see tears in Denny's eyes, something she had seldom witnessed in over thirty years of marriage. He was a bull of a man, her man—the rock of their family, capable of facing strenuous work, long hours, and arduous conditions without so much of a whisper of protest. He met one challenge in life after another with unwavering determination. He had overcome flooding, droughts, crop failures, tight financial times, stillborn calves, sick and dying livestock, and many other obstacles. They had been through rough times so often, but his tears suggested a much deeper crisis. She quickly again grabbed him in her arms to hold him tight, reassuring him that no matter how bad, "Things are gonna be all right. If we've got each other and our family, we'll get through this, Den."

"Not this time, honey," Denny confessed. "If the bank won't increase our operating loan, we are shit out of luck. We are behind in payments to the co-op, and they are threatening to cut us off. My pickup needs work, our equipment is falling down around us, and we are one of the last farms in the valley trying to make it with a small dairy herd. I don't see no daylight here!"

"Have you asked the bank to increase our loan?" Denny nodded his head. "When do you expect an answer from the bank, Denny?" gently inquired Mary.

"Probably sometime later today, but I'm not hopeful. Hell, I might as well exchange the Century Farm sign to For Sale now and avoid the embarrassment of having to talk with Tom down at the bank."

"Let's not do anything rash, Denny. I have tomorrow, Memorial Day, off. The kids are busy with their plans, giving us plenty of time to talk over our options. Promise me you'll keep your chin up," said a worried Mary.

"Sure," he said unconvincingly as he let go of their embrace. Eager for a distraction to move on from this discussion, he walked back over to the old Zenith and switched it on. Ironically, the local country music station was playing Brooks and Dunn's "Hard Working Man."

"I'm a hard, hard workin' man

I got it all on the line

For a peace of the promised land

I'm burnin' my candle at both ends

'Bout the only way to keep the fire goin'

Is to outrun the wind."

"That's it," thought Denny. "I feel like I'm trying to outrun the wind. But this is one foot race I'm afraid I'm destined to lose." More than anything, he dreaded the answer to the question of "What's next?"

Twice a day, at 4 a.m. and 4 p.m., ninety-one-year-old Emily Hanzel helps her son Dave milk his herd of twenty-five Holsteins. Dave is one of the last small dairy farmers in the Upper Pine River Valley. Copyright 2020 John P. Gauder.

The Hanzel barn, a fine example of the iconic Wisconsin family dairy barn. Copyright 2020 John P. Gauder.

Mike (right) and Carlyle Peckham relaxing in the "office" of their family farm. Mike was one of the last dairy farmers in the West Pine River Valley. Copyright 2020 John P. Gauder.

Memories of Work and Place

Later that morning, after completing milking, with Mary tending to her other usual chore of looking after the heifers, Denny jumped on the old John Deere tractor, pulling the baler and hay wagon to bring in the first crop of cut hay drying in the field along the river. As he bumped down the tractor path, he was reminded of the leaky hydraulics, bad brakes, and other overdue maintenance needed on the tractor, things he had put off for way too long because of the tight money situation. Heck, the tractor was about forty-five years old. He remembered the day in 1975 when his father had it delivered. He was twelve years old and had never seen his dad prouder. This purchase was one of the first of the new generation of Deere 4430s to show up in the valley with her enclosed cab, 139 horsepower, and two-wheel drive—the best of the new breed of powerful Deere tractors. He vaguely remembered his dad telling him that it cost about twenty grand new. Hell, you'd have a hard time getting a good Deere garden tractor for that money today. His dad had immediately scooped up both him and his sister, set them on his lap, and headed down this very path to the river before returning to give his mom a similar test drive. The memory brought one of the first smiles to his face in many days.

As he turned off the path onto the field, he startled a yearling deer out of the thicket along the fence line, probably still a bit disoriented by being separated from her mother after the birth of a new spring fawn. Denny again noted the growth of the cattails as the margins of the field got closer to the West Pine River. The scientists were calling it climate change or global warming. He didn't know if that was the cause or not. What he did know was this valley had seen way more than its share of gulley-washing rainstorms during the recent spring seasons and unprecedented river flooding, causing an accelerated transition of his lower-lying fields to cattail marsh. No longer willing to gamble on losing a corn crop, he was using his lower fields for hay, and, in a change of weather patterns, the lack of excessive precipitation this spring had allowed him to get into the fields early.

As he engaged the power takeoff for the baler and started down the first windrow of raked hay, he deeply inhaled the wonderful, rich scent of fresh-cut hay and looked skyward to enjoy the warmth of sunshine on a perfect spring day. At an early age, he had settled into the rhythm of dairy farming—spring planting of crops, summer haying, fall harvesting, and the constant task of the twice-daily milking of the herd. Some might complain that the demands of this life left little time for leisure, but Denny thanked his lucky stars that his chosen occupation allowed him the opportunity to work outdoors in this glorious Driftless landscape. Today, baling hay was akin to leisure. He marveled that a job that took several individuals a few years back could be accomplished today by himself. As the first rectangular bale came out of the baler, it was automatically tossed in the air to land squarely in the old hay wagon.

As he continued his work, his mind flashbacked to his youth when bringing in the hay was a family affair. His grandpa, a worn old man in his late seventies, with an old straw hat, a mouth full of chaw (with very few teeth), came out from the city to drive the tractor, trailing the hay wagon slowly down the field between the bales. Denny and his sister would run around the field to pick up the bales and bring them to the wagon where their dad would expertly hoist and stack them onto the wagon in such a way as to balance the load to avoid overturning the wagon, something, unfortunately, Denny had not mastered the first time his dad let him try the job of stacking. He remembered thinking during these times that his dad was probably the strongest man in the world, the way he was able to toss around those fifty- to sixty-pound bales. Invariably, Grandpa would fall asleep at the wheel of the tractor in the hot summer sun, run over a bale and damn near turn over the tractor and hay wagon, causing my dad to curse him loudly to wake up and drive right. The memories reminded him of the great place this farm was for a young boy to grow up before sadly acknowledging that the next generation of Schulz children would not be so fortunate.

As Denny returned to the barnyard with a full hay wagon, he noted that Mary had already left for her shift at the home. Deciding to wait until later to have Sam, the hired hand, help him move the bales into the haymow, he parked the tractor near the elevator and walked up to the house, letting the old screen door slam behind him as he made his way to the fridge. Not feeling much like eating, he poured himself a good strong Jack and Coke in an insulated mug. Making sure his cellphone was in the back pocket of his blue jeans in case the bank called, he decided to take a quick walk down his farm lane into what was left of the village of Woodstock.

Strolling down Schulz Lane, he was reminded of how overgrown the fence rows had become. Once those invasives—buckthorn, prickly ash, and multiflora rose—become established, it is darned well impossible to get rid of them. With their thorns and sharp spines, trying to control them became a painful nightmare. Mary, ever the one to pay attention to the condition of the landscape around the farm, kept urging him to get a handle on these unwanted intruders. He tried to beg off this miserable job by suggesting that a brushy fence row was an important environment for birds and other wildlife species, and, so far, he was succeeding with this argument.

As he turned off the lane, a street sign announced that he had entered South Shore Drive, ironic in that the street signs of Woodstock continued to stand as proud sentinels of a village with few residents and fewer homes. He paused for some time to take a long draw on his beverage, looking at a lot along the river that was rich in Schulz family lore. The story went that this was the site of the original Schulz family farmhouse, a log home cursed with multiple floods by the often quickly rising West Pine River, a structure supposedly loathed by the first Schulz farmwife on this land, Henrietta, because of its drafty and damp interior. The greater insult was that she and her husband, Douglas, had left a comfortable homestead in Vermont to make the arduous trip west to this wild, god-forsaken land.

However, the real tragedy was to occur several years later. Late at night in the winter of 1894, Denny's great-great-grandfather, Douglas Schulz, completed milking before traveling by horseback to P.J. Shaffer's country store in nearby Bloom City. There he joined fellow farmers from the valley in a game of euchre, probably washed down with ample shots of homemade brandy. That evening a fire engulfed the log home. Speculation was that the source of the fire was the wood-burning, pot-bellied stove used to heat the drafty structure. Sadly, Douglas returned home to find that his wife and two young children had perished in the fire. Their small headstones still occupy a prominent place among the many Schulz family memorials in the Woodstock Cemetery. Within a year, the devastated Douglas built the larger, present brick Schulz family home up the lane on the hillside away from the river, remarried a much younger woman, and proceeded to have a second family to carry on the Schulz family line. Today, other than the headstones in the cemetery, the only memory of the long deceased family members was a tattered black and white formal portrait in his grandmother's photo album of a proud Douglas, his wife Henrietta and their young son and infant daughter.

Denny glanced across the river to the east at the burned-out ruins of the Methodist Church. The site of many Schulz family weddings, stories abounded of mid-afternoon ceremonies held in the structure. Following the weddings, the guests left for their home farms to complete the evening milking before returning to a cleaned-out machine shed on the Schulz farm for a raucous evening celebration of drinking and dancing. It was on one of these evening occasions where Denny remembered being introduced to his first shot of brandy, washed down by a beer, that gave him the courage to ask a cute young girl named Mary from a neighboring farm to join him for a polka. The small church sat vacant during much of Denny's lifetime and was finally consumed by fire during the winter of 2020. The few Methodists left in the valley now worshipped in a pretty brick church setting on a south facing hillside in nearby Bloom City, safely elevated from the floodwaters of the nearby West Pine River.

As Denny continued his stroll down Water Street, he passed fire number 22257, now nothing but an overgrown lot, but formerly the site of the Woodstock store, the last standing business in town that closed sometime before he was born. As a kid, Denny loved to hear stories of how his father pedaled his bike down the lane with a nickel in his pocket to choose from what was to a young kid, a mind-boggling choice of penny candy on display in a glass case at the front of the store. He smiled, remembering that his father never missed an opportunity to reinforce the message of the value of hard work to him and then his children, impressing upon them the hours spent gathering eggs from the henhouse or slopping hogs to earn those nickels.

Walking by the sign announcing Main Street, he paused at the corner of South Shore Drive and Woodstock Road to admire the large, well-kept Greek Revival style home on a rise, protected from a frequently flooding river. This was formerly the home of the best-known citizen of Woodstock, Earl Sugden, also known as the "Bard of Woodstock" because of his proclivity to write poetry and dabble in art. Although Denny started elementary school just after the state began closing the one-room country schools and moving toward consolidated school districts, his local history lessons in Richland Center frequently featured the stories of Earl Sugden. History suggests that Earl quit school his first day in the first grade, was self-taught, for the most part, until the age of twenty-one, at which time he returned to school, graduated from the eighth grade in three months, and went on to finish his education at the "County Normal School." Denny's parents proudly kept a copy of the March 1941 issue of *LIFE Magazine* in the parlor, which featured a story of Earl and his art.

Denny waved to John, the present owner and thoughtful steward of the home, but then stopped in trepidation as he felt the vibrator of his cell phone in his rear jean pocket announcing an incoming call. Glancing at the display, as he surmised, the call was from Tom down at Center Bank, most likely with the verdict on his loan request. Not usually prone to panic, Denny felt his breath and pulse quicken as he answered the phone and greeted Tom.

"Denny, I'm sorry to say that I am not able to approve your application for an addition to your operating loan ..." The rest of the call involved some expressed pleasantries and well wishes, but Denny's mind had already moved on to his last option. He knew that his next call would have to be to Pete Hammes at Broad Valley Dairy.

The former home of Earl Sugden, the "Bard of Woodstock."
Copyright 2020 John P. Gauder.

A remnant structure in the village of Woodstock that has seen better days. Copyright 2020 John P. Gauder.

Memorial Day at Woodstock Cemetery

Memorial Day dawned in a chilly, foggy drizzle. After completing the milking, Denny and Mary drove across the valley to the Woodstock Cemetery. Mary was feeling uneasy. Their halting conversation around concerns over the number of family picnics and outings dampened by the weather belied their mutual fears about the future of their dairy farm. However, a few raindrops would not stop their annual pilgrimage to the cemetery to fill the flowerpots near the Schulz family memorials.

As Denny pulled into the cemetery, Mary marveled at the well-maintained lawn, and as always, admired the tranquil setting on the west-facing hillside in the West Pine River Valley. She pointed to some of the recently refurbished headstones of the early valley settlers. A generous resident down the valley was committing his time, energy, and financial resources to uprighting and restoring fallen headstones.

Getting out of the pickup, they both circled back to drop the tailgate and carry the boxes of potted red geraniums up to the Schulz gravesites. Mary was insistent upon real geraniums, forgoing the option of the more abundant plastic plantings that adorned the majority of the gravesites in the cemetery. Denny annually tried to steer her toward the colorful imitations, protesting that they were setting themselves up for a summer of watering with the real flowers.

"It's the least we can do to honor your family, Den, and it certainly doesn't hurt us to stop by to pay our respects periodically," repeated Mary, signaling an end to that discussion.

As they stepped up to the most recent headstones in the family plot, the memorials for Denny's mother and father, they held hands and bowed their heads in quiet remembrance and prayer.

"I miss 'em, Mary," Denny quietly said. "I could sure use some of Dad's wisdom and advice about the pickle we're in." While acknowledging that his dad farmed during better days for small family dairies, they both were sure that the largely unschooled former patriarch of the family would have some guidance for them on navigating the difficult times they found themselves in.

Denny was cut from the same cloth as his dad, thought Mary. His father was born to be a dairy farmer and, like Denny, was the remarkable type of person who thrived on the strenuous life of farming. But she had watched farming take a toll on his father. From a strong bull of a man through mid-life, the rigors of dairy farming reduced him to an arthritic, bull-legged, stooped, skeletal figure in later life.

A man who simply looked "used up" by the time they buried him in his mid-seventies.

As Denny turned toward Mary, he asked, "Do you think selling the herd and getting out of the business would save me from becoming a shell of my former self in old age like Dad?"

Although they had avoided this discussion earlier in the day, Mary knew it was time to confront the issue between them. "Did we get an answer from the bank, Den?" she asked.

Without looking her in the eye, he answered, "Yep, and as we suspected, it's not good news." Pulling Mary into him with an arm around her waist, he said, "Mar, we are one of the last small dairies still struggling to make it in this valley, and I think it's time to throw in the towel. Trying to tough out these low milk prices is just digging us deeper in the hole. Besides, without additional cash from the bank, we can't sustain the place."

Wanting to proceed delicately, yet concerned over the resignation in Denny's voice, Mary quietly asked, "What's next, Denny?"

"After hearing from the bank, I talked with Pete Hammes up to Broad Valley Dairy. He's applied for a permit for a major expansion of his operation and is willing to pay an honest price for our herd and heifers. I think it is our only way out, Mar."

Denny turned toward the gravestones, shoulders sagged under the weight of disappointment, and in a tearful, soft voice simply said, "I'm sorry, Mom and Dad, but I gave it all I could."

Mary was alarmed at the burden of defeat in his voice and an undertone of despair that was, until now, completely unlike the man she shared her life with. She had never seen him like this and was actually worried about his well-being for the first time in their marriage. Sensing the gravity of the situation, she desperately wanted him to feel that this turn of events, no matter how serious, did not in any way diminish her love and admiration for him.

Standing up straight in front of him and looking him directly in the eye, she exclaimed, "Denny Schulz, I love you more than ever, and no matter what happens to our farm, I will be at your side helping you all the way. Don't you ever forget how much your family and I love you and will support you. If we have to sell the herd, so be it. Tonight, let's sit down together and go over the numbers and figure out our next steps. But for now, let's get to planting these geraniums so we can get the hell out of this rain."

Maybe it was just the setting, combined with the dismal weather, but Mary saw a hesitancy in Denny's step completely out of character with his normal robust and vigorous self, further contributing to her uneasiness about her husband's state of mind.

The Woodstock Cemetery—the final resting place of approximately seven hundred former residents of the Woodstock area. Photo by Myrt Sieger.

Goodbye to the Girls

(Three weeks later)

Denny didn't need an alarm to wake for the morning milking. Despite an abundance of sedative in Jack Daniels the night before, he had tossed and turned all night, barely sleeping a wink. It seemed that every time he dozed off, he awoke startled and in a heavy sweat due to the same nightmare. He dreamt that as the farm disappeared around him, he found himself falling into a deep black hole and, try as he may while groping around in the dark, he could find no way out. After several such episodes, he settled into a gauzy wakefulness preferable to the seesaw of the dream/panic sequence evoked by the terror of clawing at the sides of a hole looking for an escape.

He knew the upcoming day would be a tough one that he would simply have to muscle his way through. He could put a good face on anything for several hours, but he had become hung up on the question "What's next?" He and Mary had agreed the past Memorial Day evening that their best option was to sell the herd to Pete Hammes and get the hell out of dairy farming. The Broad Valley vet had been out to examine the herd and check out their overall health status. Mary would continue her job at the nursing home without the additional burden of morning milking and caring for the heifers. They could remain in their home, on the family land, and lease out the tillable acreage for corn or soybeans and make a paltry amount selling hay. Denny promised to investigate other ways to augment their income, such as enrolling in the managed forest program to save on taxes, selling a few acres on the bluff side for hunting rights and maybe even finishing beef cattle. But despite all of this, he would still need to get a job outside the farm to get along. While Mary expressed confidence that the right job opportunity would emerge for him, he, on the other hand, was not optimistic. The disintegration of the family dairy farm and the cratering of the rural farm economy had left small-town Wisconsin with very few good job options. The specter of a near minimum wage job in one of the local food processing plants or a factory job at Allen Bradley in Richland Center left him desolate—feelings akin to falling in the black hole that had become the scourge of his efforts to get some sleep.

As the clock approached four-thirty, Denny crawled out of bed quietly, hoping not to wake Mary. However, forgoing their normal routine, Mary chose to rise with Denny. She had become increasingly alarmed the past few weeks over his withdrawal and seeming sense of hopelessness. Mary was not about to let him walk down to the barn unaccompanied for the last milking of his beloved herd.

After changing into their barn clothes and brewing coffee as was their morning ritual, they walked hand in hand down to the barn. The radio played a little louder this morning to drown out their pain and obviate the need for conversation, which was sure to be difficult. They could not help but notice that each of them was a bit more deliberate with this morning's routine, taking more time than normal to scratch the girls behind the ears and caress their flanks before bending over to attach the milking machines. Although careful not to let Mary see it, Denny teared up several times as he approached a couple of his favorites, Mollie and Flora, who, in addition to having especially sweet dispositions, could calve easily, having provided longer than normal service to the family milking operation. Besides the software program he had recently begun using to track herd genetics, Denny maintained a small placard on the rafter above each stanchion listing the animal's name, date of insemination, and a few other facts necessary for healthy herd management. As he looked upward in the barn, it pained him to realize that the stanchions below each name would always be empty after this morning's milking.

As they finished up and stowed the milking machines for the last time, Denny paused to look around him. "What's going to become of this old place, Mar?" he wondered as he admired the grace and style of his beloved barn. "Jeez, I don't want it to become a catch-all boneyard for all of the old used crap we have around here."

Mary came up behind him and grabbed him around the waist. "Don't you worry, Denny Schulz. You know how I ride your ass about keeping the lawn mowed and the farmyard looking nice. I am not about to allow you to let this barn fall apart like so many of the others up and down the valley," reassured Mary. "In fact, maybe now that you'll no longer be milking, I can talk you into putting a new coat of paint on her," she joked, trying to introduce a little levity into the conversation. Denny was not amused as the question of future employment remained a raw spot with him.

Unlike every other summer morning, the cows were not put back to pasture after the morning milking. They remained in their stanchions, the last dairy cows to occupy the Schulz barn awaiting their new life in an industrial dairy.

"Are you ready for this, Denny?" asked Mary. "I think I hear the trucks coming down the lane." Sure enough, as they stood at the barn door, they saw two dual-axle pickups pulling twenty-four-foot livestock trailers with the Broad Valley Dairy logo on the sides, kicking up dust as they approached the barnyard. Denny marveled at the equipment, knowing that this was just a small sample of the hardware it took to run Pete Hammes' large operation.

As the first truck pulled up to a stop in front of the barn, Denny recognized the older man who got out of the cab as George Harriman, the former dairyman who had sold his herd and had gone to work as the foreman for Pete's dairy. Despite his arthritic gait, he briskly walked up to shake Denny's hand and gave a nod to Mary.

"Denny, Mary, it's been a while since I've been to your place, and I'm really sorry for the reason I showed up today," George said, knowing from personal experience the pain they were going through. "It was about ten years ago that I sold my herd and closed up shop, but I remember it as one of the worst days of my life. I can honestly say Pete saved my life by pulling me out of my funk and offering me the chance to get back into the dairy barn—the only job I really knew."

"Well, then, I guess you could say that Pete has bailed us both out, George. If he hadn't been willing to buy the herd, we'd be loading a much larger truck on its way to Green Bay to a sad ending for these girls," answered Denny.

"By the way, Denny, Pete sends his regrets that he couldn't come today. He had to be in Madison on business, but you can reach him on his cellphone in case anything comes up," explained George.

Not wanting to get into a discussion of where he was going next and, as usual, wanting to avoid small talk, Denny was eager to get on with the job of loading up the animals.

"We might as well get started, George. Judging by the size of your trailers, it looks like you will be making several trips up and down the valley. I've rigged up a chute from the barn door with fence panels. I'll help you back up to it so you can drop the ramp. As you probably suspect, these animals are not used to being loaded onto a trailer, so this is not going to be particularly easy."

Denny's words were prophetic. The first eleven cows to be loaded into the trailer stubbornly resisted, noisily and obstinately planting their hooves and refusing to move up the ramp into the unfamiliar environment. A halter was used to lead the animals that would tolerate it, and a combination of grain and molasses was supplemented to coax the always hungry animals forward.

It was painful for Denny to endure this process. The frightened look in his cows' eyes combined with their higher-pitched bellowing was a clear sign of their fear and discomfort. Instead of their normal docile selves, they were jumpy and distraught. It made him wonder whether it was not only the unfamiliar procedure causing their agitation but also the realization that they were leaving their familiar home on the Schulz family farm forever, leading to a cry for help. Denny was sure Mary was having similar thoughts as she watched their beloved animals being loaded into the trailers for a ride to their new dairy.

Several hours later, after the frustrating job of loading the last of the herd onto the trailer for transport to the Broad Valley Dairy, Denny sat at the kitchen table in the old farmhouse, quietly downing his first Jack and coke of the day. Mary slowly walked down the stairs from the bedroom, dressed in her whites for the shift at the home and joined him at the table. She was more than a little concerned over his drinking so early in the day but did not want to come across as a nag after what surely had to be one of the most difficult mornings of his life.

Reaching across the table and grabbing his hand, she sighed, "Well, it's over, Den. Are you doing OK?"

Looking up at Mary, he quietly said, "So what do I know, Mar? Twice a day, we milked that herd. And every other chore—haying, planting, harvesting, fixing equipment—all done to support that herd. Now it's gone. What do I have left to do?"

"We've been through this a thousand times, Den. You've not had time off or a real vacation since our honeymoon thirty-three years ago. Start slowly by doing some things you enjoy and haven't had time to do. Work on your dad's tractor, join some of your friends for euchre night at the pub, blow the dust off your fishing pole and bring home some trout from the river. You've complained about a ton of things around this place that you have not had the time to do. And I guarantee you, once word gets around this county that you're looking for work, the opportunities will be knocking at your door."

Not wanting to signal to Mary the depths of his distress, he quietly agreed. "I suppose you're right, hon. I guess I could start by visiting some of the larger operations down the valley to find a buyer for our hay." Looking up at the old grandmother's clock on the wall, he reminded Mary, "Now, if you're not going to be late for work, you better hit the road."

Mary was reluctant to leave Denny in this state of mind. Working in the healthcare field, she was acutely aware of the benefits that counseling can provide when people find themselves in a hole. But she suspected that no matter how low Denny got, he would never call the county crisis line because it was staffed by people who might know him. He didn't want his weakness on display for the locals. She also read that the State Ag Department had recently set up a twenty-four-hour counseling line at the Farm Center, so she decided to take another tack without using the "counseling" word.

"Den, I'm leaving you the number of a new service available through State Ag. It's at the Farm Center. I've read that they can provide financial guidance for farmers experiencing tough times like ourselves. I think you should call them. I'm sure they would have advice for dairy farmers getting out of the business. And it's in Madison, where nobody knows you. What do you think?"

Wanting to move on from this topic, Denny said, "Thanks, Mar, but don't you worry about me. I'll take a look at it, and maybe I'll give them a call later this afternoon. Now you better get going."

"Love you, Den," she said as she kissed him gently on the cheek, picked up her keys, and headed out the door, not at all convinced that he was sincere about making that call.

Endings

As Denny slowly finished his Jack and Coke while ignoring the scrap of paper containing the phone number, he took a long look around the old farmhouse. He peered into the parlor and set his gaze on his grandpa's Winchester 94 mounted over the mantel of the fireplace. His father and grandfather had used this rifle to teach him the basics of marksmanship and firearm safety. Likewise, he had done the same with his son, Charlie, and daughter, Kate. It had become the go-to rifle for the annual fall ritual of the Schulz family deer hunt and had brought down more deer on the Schulz farm than anyone thought to count. In addition to the big buck mount hanging from the wall over the rifle and fireplace, it was responsible for the racks and antlers gracing nearly every outbuilding around the farmyard.

Denny slowly rose and walked over to the fireplace and took the rifle down from its pedestal. It smelled of bore solvent and oil, and he gently dusted off the smooth walnut stock. On the way out the back door, he stopped at the old cabinet and pulled out a couple of .30-30 cartridges, putting them into the pocket of his blue jeans. The screen door slammed behind him.

It was a beautiful, sunny June afternoon, getting warm but lacking the humidity that summer would eventually bring. He decided to take a walk through the hillside pasture and up the east-facing bluff. Letting himself through the gate, as always, he turned to close it behind him out of habit. He stopped short, thinking, "What the hell am I doing this for? The proverbial cows are out of the barn." So, he left the gate swinging in the breeze.

Hiking up the hillside on the stones of the well-worn cow path, he started reminiscing about things along the way that he hadn't thought about for ages. On the uphill side of the path stood the old Bur oak that had somehow avoided harvesting by the earliest loggers in this valley. He stopped to admire the magnificent tree, which stood alone in the well-grazed pasture with its deeply furrowed bark and its strong, horizontally spreading branches. He remembered summer family picnics under this tree that was probably well over two hundred years old. As a young boy, he and his cousins would play a game by joining hands around the trunk. It would take more than five of them standing arm-to-arm to completely encircle it.

Continuing further up the path, he spotted the family deer stand at the tree line to provide some cover and an uninterrupted view of the pasture and adjacent hillside for the traditional fall deer hunt. He flashed back to the fall day in his youth when he helped his dad tear down the old rotting deer stand and replace it with this stout structure made of pressure-treated 4x4s that still stood today. "How many early, cold November mornings did I sit in that stand with my thermos of coffee hoping to get off a good shot at a big buck?" he thought. He couldn't remember too many seasons when the family wasn't able to fill their tags because their big chest freezer was always stocked with plenty of venison. More recently, the size of the deer herd had just exploded in this area, causing crop damage, becoming a real nuisance for gardeners, and instilling terror in drivers during twilight hours, as evidenced by the numerous carcasses littering the rural roadsides. Well, his family certainly did their part to control the population. The family deer hunt was a rich tradition that took precedent over all else on the calendar each year come the end of November. "Will Charlie and Kate and their families continue the tradition after I'm gone?" he sadly wondered as he continued the walk up the hillside.

At the tree line marking the end of the pasture, he opened the gate to the old tractor path that led to the crest of the bluff, which was more than a bit overgrown from lack of use. He used the barrel of the Winchester to part the prickly ash and buckthorn that had grown up since the time he last ran his brush hog up the hill. "How quickly things get overgrown," he thought, bringing home the reminder that his footprint on this place—his family farm—would quickly be obscured by the march of the invasive plants and the reestablishment of the forest once it was devoid of his ongoing maintenance and care. "Who will be the steward of this land when I'm gone?"

Now the hillside was getting steeper, and he felt his breath becoming a bit more labored as he continued toward the top. The open pasture was replaced by a mature, second-growth oak and maple forest. While the sunny tractor path was crowded by brush, the forest understory shaded by the canopy of dense foliage was almost tidy by comparison, marked by a ground cover of ephemerals that had experienced their spring days in the sun and now were dying back. Looking up through the forest, Denny could see his destination, the large rock outcropping that crowned the top of the bluff.

He scrambled up the steep ascent to the top, climbing up a familiar tread of rocks and handholds and emerging in the sunshine, well above the treetops.

With the sun behind him, he had a breathtaking view of the entire Pine River Valley. The West Pine meandered down the valley, amongst newly cut green hayfields, corn crops that were already knee-high, even though it was not yet July 4th, and on higher, steeper ground, pasture for the increasingly common beef cattle.

Although it had been years since he made this hike, memories of this place overwhelmed him. Its remoteness provided safety for experimentation in his early years—his first cigarette, his first six-pack of beer shared with friends, the first time he kissed Mary.

Despite what should be happy memories, Denny could not shake the overwhelming feeling of shame associated with being the one in a long line of the Schulz family members to lose the farm. Try as he may, he saw no path forward for himself. With seemingly no hope for a productive and meaningful future, he envisioned himself becoming nothing more than a useless liability for the family.

Denny sat down on the side of the cliff, his gaze settling on the old Winchester on the rock next to him. He took a cartridge from his pocket, tossed it up and down, and put it back. Picking up the rifle, he cradled it in his lap. Its heft felt like an old friend. Slowly he levered the breach open and again pulled a shell from his pocket, chambered it, and slowly levered the breach closed. Gently, he raised the muzzle up under his throat and squeezed the trigger.

The retort from the rifle shot reverberated up and down the valley. And then silence, broken only by the trumpeting of a pair of sandhill cranes slowly and majestically flying high above the valley to return to their nest in the cattail marsh along the upper reaches of the West Pine. The tragedy that had unfolded beneath them would shake the valley and community in ways that were sadly becoming all too common in farm country around the state.

Beginnings

Manny Hernandez carefully backed his new John Deere zero-turn mower off the tilt trailer and onto the turf at the Woodstock Cemetery. As his doctor had prophetically suggested, he continued to experience soreness in his ribcage, although things seemed to improve with each passing day. Despite his loyalty to Pete Hammes and the Broad Valley Dairy for providing many years of good employment and assistance in securing his green card, he had finally heeded his wife's pleadings to give up the job in the dairy. After many evenings of soul searching with Kathy and discussions with friends in the Hispanic community, he and Kathy scraped together every dollar they could, augmented their meager savings with a small business loan, and purchased enough equipment to establish a one-person landscaping LLC.

Manny had discovered a rather promising client base among the seventeen cemeteries that needed lawn maintenance within Richland County. He didn't mind the long hours and handwork associated with landscaping and immediately found enjoyment in the peaceful settings of the cemeteries. He relished the independence of working alone among the trees and headstones gracing some of the most picturesque hillsides in the county.

As this was his first-time trip to the Woodstock Cemetery, he paused on his mower to admire the well-groomed and tidy groupings of over seven hundred headstones and the overall pastoral environs of the West Pine River Valley. "Yes," he thought, "this job is going to agree with me."

As he started to mow along the lane looping through the cemetery, he couldn't help but notice a newly seeded gravesite marked by abundant plantings of fresh, red geraniums. Curious, he turned off the mower, set the brake, and walked over to take a look. He found himself surrounded by gravestones with a familiar name in these parts, Schulz, some dating as far back as the mid to late 1800s. And it immediately dawned on him that he was approaching the final resting place of Denny Schulz, the local dairy farmer whose recent suicide had shaken this largely agricultural community to its very core. His passing had prompted much soul searching within the region around the plight of the family dairy farmer and the tragedy of suicides becoming all too common among struggling farmers.

Mary Schulz, the widow Denny left behind, featured prominently in the area news since his suicide. Manny and his wife, Kathy, had marveled at the strength and bravery of Mary. Although the pain associated with her loss was clear, she had nonetheless used her grief to become a very vocal and effective advocate for statewide resources to address the crisis in the dairy industry and the spate of resulting suicides. Mary recently had been interviewed on the local AM radio station immediately after the farm report, using the occasion to promote the Richland County Crisis Counseling line and imploring state government leaders to support crisis counseling resources desperately needed at the local level. She tearfully suggested that if her husband had made "the call," he might still be alive today.

Having just come through a crisis of his own, his workplace accident at Broad Valley Dairy, Manny had a fresh appreciation for the fragility of life. He thanked his lucky stars daily that the outcome had not been more serious, realizing that he came within inches of serious internal damage and possibly even death.

Manny fell to his knees as he approached the gravesite, crossed himself, and said a silent prayer for the repose of Denny's soul and for strength and healing for the family he left behind. He ended the supplication to his God with often repeated entreaties to watch over the health and well-being of his wife, Kathy, and his family.

Climbing back onto his mower and starting the engine, Manny reflected on how the sober encounter with Denny's gravesite reinforced a spirit of thankfulness for a new beginning in life and a fresh start in his landscaping business. He paused to look skyward and feel the warmth of the summer sun on his face. His gaze landed on a pair of red-tailed hawks circling above him on the thermals over the valley. He interpreted this as a good sign because legend says that seeing hawks means you are on the right path in life.

Afterword by the Author

As a young boy growing up in La Crosse, I developed an early and lasting love of the Driftless landscape. Summer days were spent hiking with siblings and childhood friends in the bluffs framing the eastern border of the prairie along the Mississippi River. Trails led to unique and, to the young mind, mysterious rocky outcroppings on which were conferred names such as Fat Man's Misery, Indian Joe's Cave, and Grandad's Bluff. We could while away countless hours trekking along these trails, climbing to higher heights and exploring previously unvisited segments of the rugged landscape.

However, annual visits to a cousin's farm left even a greater impression on the mind of this young man. Each summer, on a much-anticipated day, Cousin Alvin would pull up to the curb in front of our family home in his old pickup to collect my grandmother and me and take us to the family farm in Vernon County, Wisconsin. Since he was coming into the city and before showing up at our family home, Alvin always swung by the old G. Heileman Brewery to pick up a load of waste malt from the brewing operation, a cheap and nutritious source of feed for his small dairy herd. His load always leaked a pungent, spoiled beer-like liquid as my grandma and I crawled into the cab for the ride to Chaseburg. Then for two glorious weeks, we were guests for, what was to me, the idyllic life on a small family farm. Bringing the cows in from pasture, helping with the twice-a-day milking, assisting in haying operations, gathering eggs from the henhouse, and hoeing the endless rows of tobacco plants were some of the chores I was tasked with. While I've come to appreciate that the work required of a small dairy farm is endless and can be backbreaking and tedious, to a young city boy, it was nothing short of magical. It left me with a lifelong love, certainly idealized, for the small family dairy farm.

As I began to compile research for this book, I became increasingly alarmed at the plight of the small family dairy farm in Wisconsin. The statistics were dismal. With rising milk prices in 2014, farmers in Wisconsin and across the nation responded by bringing more cows into milk production. This eventually led to oversupply problems. On top of that, trade wars and tariffs instigated by the Trump administration led to the loss of key dairy export markets like Canada, Mexico, Europe, and China, and prices plunged. The small dairy operation was hit the hardest. In the last decade, the U.S. lost nearly twenty thousand licensed dairy farms. Wisconsin alone has lost more than sixteen hundred dairy farms in the last three years. The setting for this book, the West Pine River Valley, has seen the number of its dairy farms go from fifty-five historically to zero today. Just as prices began to rise at the end of 2019, the world was hit with the COVID-19 pandemic in early 2020, unsettling the dairy industry and resulting in reports of milk dumping related to the loss of demand in restaurants and schools across the nation. During this time, Wisconsin led the nation in farm bankruptcies.

The dairy industry's answer to these difficult times seems to have been "get big or get out." In what has become an increasingly cut-throat commodity market, dairy farmers often have tried to reap the benefit of economy of scale by increasing the size of dairying operations. The average herd size in the U.S. has doubled to about two hundred fifty cows and, while there are fewer dairy farms, they are larger and producing more milk than ever. This all raises the unsettling specter of the rapid disappearance of the small dairy farm from the landscape of Wisconsin.

Can or should something be done to reverse or somehow stabilize this trend? Dairy producers often complain of the complex and unwieldy federal minimum-pricing formulas that the Milk Marketing Order law specifies for fluid milk that processors must pay to dairy farmers or cooperatives. A major concern is that the price, which is set by the marketing order, has nothing to do with the cost of production. It is based upon supply and demand and how far the milk is produced from Eau Claire, Wisconsin. Repeated calls for revisions to this federal law, passed in 1937, go nowhere in a federal legislature characterized by a polarized, intractable environment.

Also, the consumer demand for fluid cow's milk products is steadily decreasing. Consumers now have the choice of a variety of plant-based milks from products derived from oats, soy, rice, and almonds, among others. The consumption of these products has been rapidly increasing over the last fifteen years, replacing the demand for cow's milk and further stressing the dairy industry.

On a state level, in his 2020 State of the State address, Governor Tony Evers issued a call to the legislature to come into special session to address the plight of the small dairy farmers, saying, "We have not forgotten those who have shared the harvest and bounty, feeding our families, our communities, our state, and our country for more than a century. And tonight, we say that we are ready to be a partner in the promise of prosperity." Among the proposals was establishing five regional positions across the state to provide mental health support for struggling farmers. These are needed resources as more and more rural communities sadly deal with the tragedy of suicide among their farming friends and neighbors. Suicide rates for farmers are more than twice that of non-farmers and are certainly exacerbated in the rough years we have recently seen.

Sadly, Assembly Speaker Robin Vos, rather than rolling up his sleeves to work with the governor in the interest of farmers in Wisconsin, criticized the move, saying the governor's proposals are "a lot of rehashing of ideas he had in the budget, which really just expanded the bureaucrats in Madison and added more state positions."

This was followed by more political polarization over farmer mental health issues between the governor and the Wisconsin Legislature. Earlier, Secretary Designee of the Wisconsin Department of Agriculture, Trade and Consumer Protection, Brad Pfaff, had called on the legislature to release funding for mental health vouchers for farmers and farm families in crisis. Wisconsin Senate Majority Leader Scott Fitzgerald called the secretary designee's comments "offensive and unproductive." And in an unusual and vindictive move, the Senate refused to approve Brad Pfaff's appointment as secretary of Wisconsin's agricultural agency even though he was well regarded and had the support of the agricultural industry within the state. It appears that our state government leaders are unable to seek bipartisan solutions to a crisis facing the family dairy farm, an important institution in Wisconsin.

Still, there are some hopeful stories among the small dairy farms of the Driftless. Certainly, one of these is the success of Organic Valley, a farmer-owned cooperative founded in 1988 and located in the heart of the Driftless in La Farge, Wisconsin. Begun initially with a focus on wholesome vegetables grown without pesticides, herbicides, and commercial fertilizers, they soon added organic dairy products. Initially, they proved to be a lifeline for small dairy farmers. While regular milk consumption has remained flat since the mid-1980s, the consumer appetite for organic milk began growing steadily in the mid-1990s. A small dairy farmer agreeing to feed and manage their herd without hormones or antibiotics, allowing the herd access to pasture during the growing season, and ensuring that forage and feed crops are free of prohibited substances can move toward USDA accredited organic certification in three years. One of the major benefits of this significant undertaking is the receipt of a premium for milk production. The average price paid by the co-op to their member farms in 2019 was $30.25 per hundredweight of milk, about $12 more than conventional farmers receive. Despite the more complicated and potentially costly organic operations, this cost differential can result in a difference of around $150,000 in annual revenue to a dairyman with a herd of one hundred cows.

The initial success of Organic Valley led to its becoming the largest farmer-owned cooperative in the nation, reaching $1 billion in sales in 2016. However, even this successful venture has faced recent challenges. With decreased demand for their dairy products between 2017 and 2019 along with oversupply problems, Organic Valley lost money for the third straight year in 2019. Today, they have declared a moratorium on adding new member farms. There were three hundred Wisconsin farm members at the end of 2019, but Organic Valley is hopeful amidst signs that organic milk consumption is increasing significantly while people are staying at home during the COVID-19 pandemic.

Another successful model being embraced by a small number of dairymen is the return to the early days of the farm dairy, that is, keeping the milk on the farm for processing and direct sales of finished dairy products to the consumer or retail market. The success of Pleasant Ridge Cheeses in Dodgeville is a fine example. Business partners Dan Patenaude and Mike Gingrich use the spring and summer milk from their pasture-fed cows to craft an award-winning artisanal cheese called Pleasant Ridge Reserve. Due to hard work and investment, they have achieved success beyond their initial dreams by not selling their fluid milk to a factory cheesemaker for $17 per hundredweight. Instead, they figured that their farm dairy-cheese operation provides a return of something near to $100 per hundredweight.

The former dairy farms also are being repurposed for other agricultural uses. Entrepreneurs Josh and Noah Engels grew up on organic dairy farms and today operate Driftless Organics on one hundred acres of former dairy farmland ridges and valleys near Soldiers Grove, Wisconsin. Converting former dairy farmland into an organic cropping operation, they have taken maximum advantage of consumers' preferences for locally grown produce to successfully raise organic fruits, vegetables, and sunflower oil for sale in farm markets and grocery stores throughout the upper Midwest. With a careful and sensible land ethic, the partners' one hundred acres in the heart of the Driftless will no longer be supporting grazing dairy cows. This model is supported by advocates with an eye toward environmental conservation and sustainability. They call for a transition away from the dairy industry in the Driftless so that the hillsides and stream banks will be spared erosion due to the far too common overgrazing, and local waterways will no longer receive excessive nutrient runoff from cow manure.

The collapse of the family dairy farm industry has had severe negative consequences for the small-town economy of the Driftless. While the Schulz family featured in this story is fictional, the setting in the village of Woodstock is very real. Other than a few residents in the sparse number of remaining homes, little vestiges of this community other than memories of an earlier, more prosperous time exist today. One of those proud memories is the "Bard of Woodstock," Earl Sugden. A largely self-taught philosopher, composer, scientist, artist, musician, and composer, Earl taught for sixteen years in the schools of rural Richland County and lived in a stately Greek revival home that still stands in the village. The March 31,

1941, issue of *Life Magazine* featured a story on Earl and his creations. Earl died in 1974 at the age of 89.

Small towns and unincorporated villages of the Driftless such as Rockbridge, Hub City, Bloom City, Yuba, Cazenovia, and Woodstock are deteriorating, some seemingly almost on the verge of disappearing. And yet, even here, one can find success stories. For example, Richland Center has been the beneficiary of a hard-working family that has recently reinvigorated an old furniture store in their central business district, reminding local citizens of the benefits of shopping and purchasing locally. La Farge and the surrounding Kickapoo Valley have enjoyed recent investment by the Organic Valley Co-op and have seen increased interest in tourism focused on canoeing and paddling the scenic Kickapoo River. Viroqua has become a Driftless area destination with a newly enlarged food cooperative, a James Beard award-winning restaurant in the Driftless Cafe, as well as numerous other locally owned businesses making up its quaint, historic downtown business district.

During the development and evolution of the dairy industry, the University of Wisconsin, a crown jewel within our state, has been an invaluable resource, partner, and strong advocate for farmers. The UW–Madison College of Agriculture and Life Sciences, founded in 1889, has led the way in the industry with its introduction of the Wisconsin Dairy Barn in the late 1800s, the tower silo, the Babcock butterfat test in 1890, and Vitamin D research, to name a few early innovations. Today it remains a valuable partner with the dairy industry, translating research into practice in areas such as animal health and nutrition, genetics, marketing and economics, and much more. Most recently, the agriculture schools in the UW campuses of Madison, Platteville, and River Falls joined forces to create the UW Dairy Hub to harness the capacity of all three programs to find solutions to the complex challenges facing the industry.

On March 2, 2020, the UW–Madison sponsored an event focusing on the future of the dairy industry, which included a tour of the Mystic Valley Dairy in northern Dane County, Wisconsin. As a nostalgic aficionado of the family dairy farm, I was skeptical of what I would find in a large dairy milking over four hundred cows. However, Mystic Valley Dairy, owned and operated by Mitch Bruenig and his family, dispelled any negative perceptions I had regarding large dairying operations. The Holstein herd was beautiful and well cared for in an immaculate free-stall barn, the milking parlor was efficient and sanitary, and the employees were friendly and welcoming. Most importantly, it was evident that Mitch and his family, including a daughter with a recent degree from the UW–Madison Ag School, were proud of their dairy, cared deeply for their operation, and appeared to be thoughtful stewards of their land and dairy herd.

As small dairy farms continue to disappear from the Driftless area, I am hopeful for the future. Whether it is the organic farmers, thoughtful large dairy operators, entrepreneurs, artisans, dreamers, or others, as long as we treasure the hills and valleys of this unique place and continue to be thoughtful stewards of this precious landscape, it will continue to provide opportunities for livelihood, creativity, and recreation in a magical environment.

Bring the Cows Home

We stumble down the rocky path with glee,
kids cracking jokes, from more difficult tasks we flee.

Clenched between our teeth a stem with tassel of silk,
the last chore of the day: bring the cows home to milk.

“Come boss, come boss,” we loudly bray,
and see the docile brown creatures start to move our way.

Back to the barn where they placidly saunter,
into the appointed stanchion they magically wander.

And the milking begins to fill the large can,
the twice-daily ritual of the Wisconsin dairyman.

After turning the cows out to pasture at night,
to begin the same process tomorrow at dawn’s first light.

That was then and this is now,
a very different life for our state dairy cow.

No days are spent in our Driftless' green meadows,
often, they are sheltered year-long in factory ghettos.

Confined to the free stall, a corporate farm you envision,
the herd to be managed with industrial precision.

Sure, the genetics are better and the yields up much,
nonetheless, I must wonder, don't you miss the earlier
touch?

When the family knew each of the cows, most often by
name,
and the milking process was not a round-the-clock game.

In summer the herd was pastured and fed grass,
but how did this new dairying process come to pass?

From a family operation to a corporate endeavor,
I know the earlier days were tough and strenuous; however,

Call me old fashioned, an idealist or dreamer,
but don't you miss the days of a simpler demeanor?

Acknowledgments

Although this book is a short work, there are a number of individuals whom I would like to thank for their guidance and assistance. Dennis Blood, who has a Century Farm on the West Pine, graciously took me around the valley, pointing out sites and introducing me to individuals who proved to be very helpful in researching the area. Among these were Mike and Carlyle Peckham, who allowed us to photograph their log cabin and invited us into their "office" to share stories and perspectives of dairy farming in the valley. Jim, Emily, and Dave Hanzel graciously opened their small dairying operation to us. John Seymour allowed photography of the exterior of his home in Woodstock, the former residence of Earl Sugden. Mitch Bruenig allowed us to photograph his family's impressive operations at the Mystic Valley Dairy.

The staff of the History Room in the Brewer Public Library in Richland Center provided a wealth of information on the early history of Woodstock. Had the pandemic not interrupted my research, I could have spent many more hours in their collection, working with their knowledgeable and helpful staff. I'd like to thank those who did an early read of the book and provided very helpful feedback: my friend Steve Shober, my neighbor Joel Rodriquez, my sister-in-law Barb Zanin, and my daughters Kate and Anne.

I also received wonderful feedback and encouragement throughout my writing process from the Wednesday PLATO Writing Group, which Andy Millman skillfully facilitates.

Maryanne Haselow-Dulin did a wonderful job editing the manuscript, and Arcane Covers provided invaluable assistance in cover design. The guidance and patience of Kristin Mitchell and her colleagues at Little Creek Press were ever so helpful in bringing this project to completion. My friend and former colleague, Robin Good, directed her formidable artistic talents toward the creation of the wonderful area map. My neighbor and good friend, John Gauder, an accomplished photographer, provided the beautiful photographs that illustrate this text. I am ever so thankful for his discriminating eye that helped in choosing the appropriate subjects and capturing the scenes from the Upper Pine River Valley.

Finally, my lifelong partner and wife, Myrt, provided the initial impetus to begin this effort and encouragement and helpful feedback every step of the way. For this and for so, so much more, I am forever grateful.

About the Author

Tom Sieger and his wife Myrt are fortunate to care for a small parcel in the West Pine River Valley where they feed butterflies and birds, nurture prairie plantings and battle invasive brambles.

Printed in the United States
by Baker & Taylor Publisher Services